MICROSCOPY HANDBOOKS 43

Electron Microscopy in Microbiology

Electron Microscopy in Microbiology

M. Hoppert
Institut für Mikrobiologie und Genetik,
Georg-August Universität, Göttingen, Germany

A. Holzenburg
School of Biochemistry and Molecular Biology and School of Biology,
University of Leeds, Leeds, UK

In association with the Royal Microscopical Society

[M. H]oppert and A. Holzenburg
[Respe]ctively Institut für Mikrobiologie und Genetik, Georg-August Universität, Grisebachstr. 8, [D-37]077 Göttingen, Germany and School of Biochemistry and Molecular Biology and School [of Bi]ology, University of Leeds, Leeds LS2 9JT UK

[Publ]ished in the United States of America, its dependent territories and Canada [by a]rrangement with BIOS Scientific Publishers Ltd, 9 Newtec Place, Magdalen Road, Oxford OX4 1RE, UK

First published 1998

A CIP catalog record for this book is available from the British Library.

Library of Congress Cataloging-in-Publication Data
Hoppert, Michael.
Electron microscopy in microbiology / Michael Hoppert, Andreas Holzenburg.
p. cm.
Includes bibliographical references (p.).
ISBN 0-387-91564-8 (alk. paper)
1. Electron microscopy–Handbooks, manuals, etc. 2. Microbiology–Handbooks, manuals, etc. I. Holzenburg, Andreas. II. Title.
QR68.5.E45H66 1998
579′.028′25–dc21 98–7729
CIP

ISBN 0-387-91564-8 Springer-Verlag New York Berlin Heidelberg SPIN 19901016

Springer-Verlag New York Inc.
175 Fifth Avenue, New York
NY 10010-7858, USA

Production Editor: Andrea Bosher.
Typeset by Poole Typesetting (Wessex) Ltd, Bournemouth, UK.
Printed by Biddles Ltd, Guildford, UK.

Front cover: Flagellated cell of *Ralstonia eutropha* after unidirectional metal shadowing (see *Figure 4.13*).

Contents

Abbreviations

B52	*p*-toluenesulphonic acid
BAC	benzyldimethylalkylammonium chloride
BSA	bovine serum albumin
DAB	diaminobenzidine
ds	double-stranded
EELS	electron energy loss spectroscopy
EF	extracellular face
EFTEM	energy-filtering transmission electron microscopy
EM	electron microscopy
ES	extracellular surface
ESI	electron spectroscopic imaging
GFP	green fluorescent protein
MME 7002	hexamethylolmelamine methyl ether
OD	optical density
PAG	Protein A–gold
PBS	phosphate-buffered saline (50mM K-phosphate buffer, pH 7.0 and 0.9% w/v NaCl, if not indicated otherwise)
PEG	polyethylene glycol
PF	protoplasmic face
PHB	poly-β-hydroxybutyric acid
PML	progressive multifocal leucoencephalopathy
PS	protoplasmic surface
R bodies	refractile bacterial inclusion bodies
S	surface
SDS	sodium dodecyl sulphate
SEM	scanning electron microscopy
ss	single-stranded
TEM	transmission electron microscopy

Preface

"One studies a complex system by dissecting it . . . then tries to obtain a detailed picture of its parts by X-ray analysis and chemical studies, and an overall picture of the intact assembly by electron microscopy."
Sir Aaron Klug (Les Prix Nobel en 1982)

For many years, electron microscopy (EM) has been an indispensable tool to further our understanding of cellular and subcellular processes. The versatility of EM studies is particularly useful in the ever-expanding field of microbiology encompassing pure and applied topics reaching from agronomy to zoology, and from biotechnology to medicine.

This handbook has been written with the intention of providing a concise, yet comprehensive, hands-on guide to all those scientists who would like to obtain structural/functional information about microorganisms and their contents using EM techniques. It is also hoped that those who are not directly involved in research, but, for example, oversee R&D activities, will find in this book ideas of what can be reasonably achieved and when to consider EM studies.

Especially in the age of advanced spectroscopical methods (NMR, Fourier-transform IR, etc.), numerous X-ray crystallographical successes, and electron crystallographical and microscopical breakthroughs, all aiming at structural details at an ever-increasing resolution, there are many, perhaps most crucial clues still to come from information on intact assemblies within a physiological or near-physiological environment.

Michael Hoppert
Andreas Holzenburg

Acknowledgements

The authors would like to thank Hans R. Gelderblom, René Hermann, Reinhard Rachel, Siân Renfrey and Drs Svetla Stoylova for helpful discussions and the provision of electron micrographs, Dr Fiona H. Shepherd for proof-reading the manuscript, and Mrs Yvonne Wüstefeld, Mr Paul McPhie and Mr Adrian Hick for expert technical assistance.

We are also grateful to Prof. Dr Frank Mayer and all our present and former co-workers for essential foundations, without which this book could not have been written.

Safety

Attention to safety aspects is an integral part of all laboratory procedures, and both the Health and Safety at Work Act and the COSHH regulations impose legal requirements on those persons planning or carrying out such procedures.

In this and other Handbooks every effort has been made to ensure that the recipes, formulae and practical procedures are accurate and safe. However, it remains the responsibility of the reader to ensure that the procedures which are followed are carried out in a safe manner and that all necessary COSHH requirements have been looked up and implemented. Any specific safety instructions relating to items of laboratory equipment must also be followed.

1 Introduction

The range of electron microscopical applications in microbiology alone is vast, and can appear overwhelming for someone with little or no experience in transmission electron microscopy (TEM) or scanning electron microscopy (SEM). In this context, two excellent surveys can be recommended. One constitutes part of a volume of the Encyclopedia of Plant Anatomy (Mayer, 1986) — and deals nevertheless with bacteria — and the other is a review article by Holt and Beveridge (1982).

While some applications require special apparatus, in addition to the obligatory electron microscope and coating unit, some straightforward and versatile techniques are available, for example, negative staining. These are very useful for structural and biochemical characterization, thus complementing gel electrophoretic and immunological techniques.

The detailed descriptions of the experimental procedures presented within this book are based on experience gathered over the years in the authors' laboratories and prove to be highly effective tools for microbiologists' work.

The documentation of the numerous possible protocols and variations thereof would be far beyond the scope of a small handbook in this series. Therefore, the protocols presented in detail have been selected on the basis of their importance for studies on the structure–function relationships of microbial cells. Additional sophisticated equipment is described in detail, and special emphasis has been placed on cryopreparation techniques and immunocytochemical techniques. These latter techniques are especially important to those seeking more detailed information on the compartmentation of bacterial cells.

2 General remarks

Some of the chemicals used for electron microscopic preparation have to be of the highest quality and purity ("EM" grade), that is, all heavy metal staining salts, the reagents used for immunocytochemical localization, fixatives and all chemicals used for DNA spreading. For DNA-spreading techniques, especially, only water of the highest purity should be used. For all other preparation techniques described in this book, double-distilled water, which has been freed from contaminating particles by ultrafiltration or centrifugation, must be used.

2.1 Safety measures

Some of the chemicals used for electron microscopic preparation are (very) toxic, radioactive, carcinogenic or teratogenic or consist of volatile hazardous compounds. Before attempting any of the following procedures, it is, therefore, essential to consider appropriate safety measures carefully.

A detailed description of hazardous chemicals mentioned in this book is given in Appendix D.

3 Grids and specimen support films

The standard diameter of specimen support grids used in TEM is 3.05 mm. Usually, the grid has to be covered with an electron-translucent specimen support film prior to use (*Tables 3.1* and *3.2*). The grid type should be selected with respect to the specimen and the preparation; some examples are listed in *Table 3.1*. The stability of the specimen and the size of the field of view are two parameters which have to be taken into consideration when making the choice. For most preparations, copper grids provide sufficient stability. Nickel, gold or gold-coated copper grids are recommended for procedures involving the application of copper-oxidizing chemicals.

Grids with small square or hexagonal meshes are most frequently used for standard preparations. Grids with single holes or slots exhibit a maximum field of view, but provide little specimen support. With the so-called "finder" grids, individually labelled meshes facilitate the relocation of interesting areas of the specimen.

On the one hand, specimen support films should provide optimal stabilization of the specimen during preparation and in the electron microscope, while on the other, minimal interaction with the incident electron beam is required in order to obtain images with a low background noise. Therefore, thin ($\leq$ 20 nm thick) structurally homogenous support films of maximum strength should be used. One also needs to consider whether the surface properties of the support film (e.g. its hydrophilicity) facilitate efficient adhesion of the sample.

3.1 Plastic support films

Plastic films are decomposed by the electron beam and tend to drift during the first seconds of irradiation. Nevertheless, they usually provide sufficient stabilization for ultra-thin sections of resin-embedded samples (see below). Carbon coating (see below) stabilizes the structure

Table 3.1. Grids and specimen support films for different specimen preparations

Specimen	Support film	Preparation/sample characteristics	Grid specifications
Protein molecules, subcellular structures, bacterial cells	Carbon film (very thin carbon films need to be supported by a holey plastic film)	Negative staining, unidirectional metal shadowing	Small square or hexagonal mesh (700–400 mesh in^{-1}) copper
	Holey carbon, which has been rendered hydrophilic	Frozen hydrated specimens	Small square or hexagonal mesh (700–400 mesh in^{-1}) copper
Nucleic acids	Collodion	Rotary metal shadowing	Small square or hexagonal mesh (700–400 mesh in^{-1}) copper
Bacterial cells and cellular aggregates	Formvar	Resin embedding, ultra-thin sections (70–80 nm thick)	
		• Small specimens (e.g. single bacterial cells)	Square/hexagonal mesh (400 mesh in^{-1}) copper
		• Large specimens (e.g. large aggregates or tissue samples)	100–300 mesh/in^{-1} copper
		• Ribbons of serial sections	Single slot, multiple slots copper
		• Subsequent treatment (e.g. immunocyto-chemical localization)	Nickel
	No support film	Very thin sections (40–50 nm thick) for elemental analysis	Small square/hexagonal mesh (700–400 mesh in^{-1}) copper
	Formvar, carbon-coated Formvar	Cryosections	Square/hexagonal mesh (400 mesh in^{-1}) copper
Large specimens with small areas of interest	No support film	Specimens for SEM	Finder grids of different form and dimension

of plastic films. Generally, plastic films are not suitable for the visualization of specimens at high resolution. In *Table 3.2*, some widely used types of plastic support films are listed. Though Formvar and carbon-coated Formvar are the most widely used support films for TEM, numerous other plastic materials are suitable, especially when the physical properties of the support film (and, therefore, the interaction

Table 3.2. Examples of plastic support films

Film	Preparation	Useful as support film for
Formvar	Film prepared from polyvinylformaldehyde 0.5% (w/v) in chloroform (Drummond, 1950)	Ultra-thin sections of resin-embedded specimens
Carbon-coated Formvar	Formvar-coated grids coated with carbon (Bradley, 1954)	Cryosections; "whole-mount" preparations (see below)
Collodion	Prepared from a 2% (w/v) solution in isoamyl acetate (Drummond, 1950)	DNA molecules (visualized by subsequent rotary metal shadowing)
Carbon-coated collodion	Collodion-coated grids additionally coated with carbon (Bradley, 1954)	Membranes and membrane proteins
Holey carbon	Holey collodion film (prepared from 0.2%, w/v, collodion in ethyl acetate) coated with carbon (Lünsdorf and Spiess, 1987)	Very thin carbon films in high-resolution EM of negatively stained or frozen hydrated specimens (see below)

between the support and the sample) need to be varied. Other support films are based, for example, on cellulose acetate (solvent: acetone, ethyl acetate), methyl methacrylate (solvent: chloroform) or polystyrene (solvent: benzene). In *Tables 3.3* and *3.4*, accompanied by *Figure 3.1*, the experimental procedures for production of support films are described in detail.

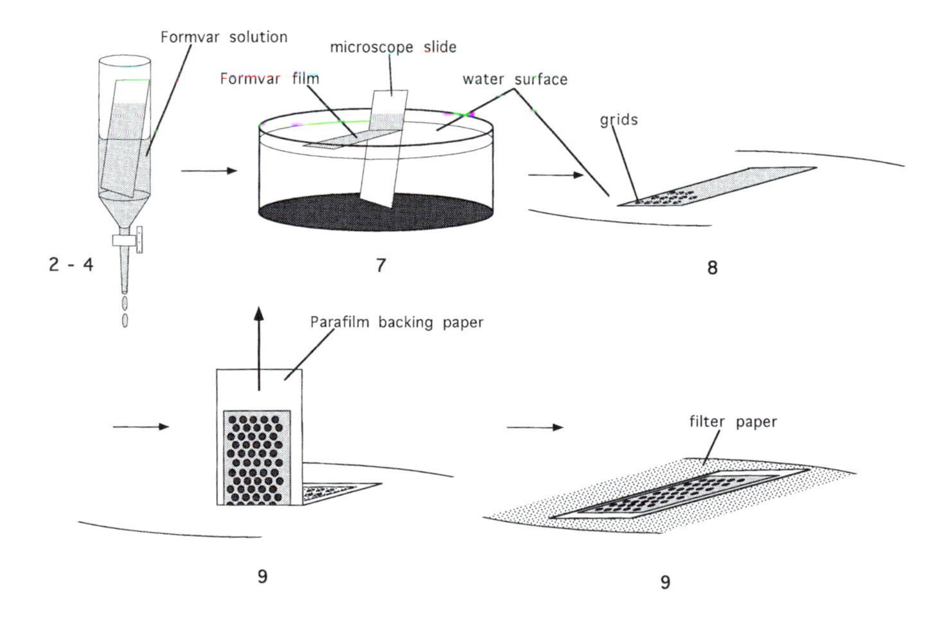

Figure 3.1. Preparation of Formvar films. The figures refer to preparation steps given in *Table 3.4*.

Table 3.3. Preparation of collodion film by casting on water

1. Clean a glass Petri dish/Büchner funnel fitted with a tap, rinse several times with double-distilled water and then fill with double-distilled water
2. Place a fine mesh wire gauze or a piece of filter paper (latter is sufficient when using a Büchner funnel) inside
3. Place the grids (shiny side up or down according to your own convention, but bear in mind that the shiny side has a larger contact surface area) on to the gauze/filter paper
4. Using capillary forces only, fill the tip of a long Pasteur pipette with a solution of 2% (w/v) collodion in (water-free!) isoamyl acetate. Then allow one drop of this solution to fall on to the clean water surface (illuminating the surface at an angle of 45° reveals major contaminants — these can be removed with the aid of a fluff-free filter paper) at the centre of the dish/funnel from a height of about 5 cm
5. Allow the solvent to evaporate during which interference colours can be observed on the nascent film. Once interference has stopped, wrinkles at the periphery of the pale grey film indicate that all the solvent has evaporated and that the film is ready to be lowered on to the grids underneath by slowly draining off the water
6. Place the gauze/filter paper on a piece of filter paper with a larger diameter, provide it with a cover, and allow the collodion-coated grids to dry (do not place them in a hot oven)

Table 3.4. Preparation of Formvar films by casting on glass (*Figure 3.1*)

1. Clean a microscope slide using detergent and water. When drying the slide, a thin residual layer of detergent remains on the surface of the slide to facilitate step 6
2. Place the slide inside a cylindrical glass funnel (diameter: 3.2 cm; length: 8 cm; fitted with a tap and drain tube) which has been filled with a 0.5% (w/v) solution of Formvar in water-free (!) chloroform or ethylene dichloride. The slide should not be totally submerged but by a maximum of 3/4 of its length
3. Cover the funnel with a lid
4. After ~30 sec, open the tap and let the solution drain off
5. After a further 30 sec, remove the slide using forceps and allow it to dry
6. Scratch the edges of the glass slide with a razor blade and breathe on the film prior to immersion
7. Float the film off onto a clean water surface. This is achieved by slowly lowering the short edge of the slide into a dish filled with double-distilled water (almost up to its rim) at an angle of 30–45°
8. Place grids on top of the film, starting at each of the four corners in order to facilitate visualization of the area to be covered
9. A suitably sized piece of Parafilm™ *backing* paper (not the Parafilm itself!) is now lowered on to the grids and gently pressed home. Using forceps, the whole "sandwich" is then removed from the water surface, turned upside-down, placed on a piece of filter paper, and left to dry overnight

The thickness of the film is essentially dependent on two parameters, the concentration of the Formvar solution and the time allowed for the solution to drain off.

3.2 Carbon and carbon-coated plastic support films

Carbon is the most widely used film for the direct adsorption of small specimens such as protein molecules. Very thin carbon films may be supported by a holey film (see *Table 3.2*). Carbon films are produced by resistance evaporation and subsequent sublimation of carbon on to, for

Table 3.5. Preparation of holey films

1. Wash 0.1 g collodion with ethanol, allow to dry, and dissolve overnight in 75 ml of ethanol in a brown glass-stoppered flask at room temperature
2. Add 125 μl of 10% (w/v in water) BRIJ 58 (other detergents or fabric softeners may do equally, but optimization may be required) and 0.92 g of 87% (v/v in water) glycerol
3. Shake gently for 2–3 min (shaking too vigorously results in too small holes)
4. Transfer the turbid solution into a beaker and dip glass slides into the solution for 10 sec. Retract them slowly and allow to dry in an upright position inside a clean beaker for 10 min
5. Proceed as for Formvar films (Task 3.4, step 6 onwards)
6. Bake the grids for 10 min at 180°C
7. Stabilize the holey films by coating them with carbon

example, Formvar- or collodion-coated grids, holey grids (see *Table 3.5*) or freshly cleaved mica as an intermediate carrier (see also Kölbel, 1976). In the last case, carbon-only support films are produced. In order to achieve a uniform distribution of small carbon particles and improved smoothness, the surface can be prepared by *indirect* carbon coating, during which the evaporated carbon is reflected from a glass slide at an angle of 45° to the target surface (mica). Sublimation of non-reflected carbon (direct coating) is prevented by a metal shield between the evaporation source and the mica (*Figure 3.2*). The carbon film is subsequently floated off on to a clean water surface and grids are placed on top of the film. Mounting is completed as in the method detailed for Formvar films.

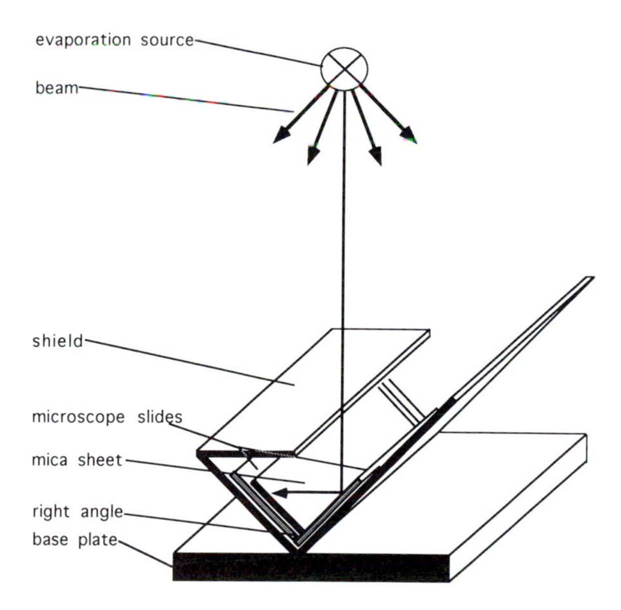

Figure 3.2. Indirect carbon coating. The simple apparatus, preformed from sheet steel, is fixed on a metal base plate. A shield prevents the mica plate from direct exposure to the carbon beam. The beam is deflected as indicated by the large arrow. Redrawn from Robinson *et al.* (1987).

3.2.1 Preparation of hydrophilic films by glow discharging

In order to increase the hydrophilicity of the support film surface, grids may be glow discharged immediately prior to use (hydrophilicity of the grid surface is stable for about half a day) under reduced atmospheric pressure (~ 0.1 Torr) in air (resulting surface charge is negative) or in the presence of alkylamines (resulting surface charge is positive). Glow discharging improves sample adhesion and spreading of the staining solutions (Dubochet *et al.*, 1971). The use of hydrophilic support films is essential for the preparation of frozen-hydrated samples and extremely useful when preparing macromolecular structures and membranes.

3.3 Other support films

For high-resolution imaging of specimens, aluminium oxide films with minimal intrinsic structure (Müller *et al.*, 1979) and very thin carbon films ("mono-films", Sakata *et al.*, 1991) have been developed. The preparation of these support films is very elaborate and/or requires special equipment and is, therefore, only of limited use to the microbiologist. Silicon monoxide films (Drummond, 1950) cannot be generally recommended as a substitute for carbon films because of their inferior stability and the fact that they are barely visible to the naked eye.

4 Preparation techniques for transmission electron microscopy (TEM)

4.1 Negative staining

In TEM, contrast is generated by the differences in interactions of the electrons with the specimen. Since biological specimens generally consist of elements of low atomic mass which interact with electrons only very weakly, contrast is conventionally imparted to the final image by the addition of heavy metals. In contrast, the specimen support film needs to be as "electron neutral" as possible in order not to introduce significant levels of background noise to the image.

Negative staining as a method of contrast enhancement in electron microscopy was first applied to the visualization of virus particles (see, e.g. Wurtz, 1992 for review). The technique is convenient, extremely rapid, and applicable to objects ranging in size from macromolecules (several nanometres; see *Figure 4.1*) to subcellular structures and whole organisms such as bacteria (several micrometres; see *Figure 4.2*). Staining is achieved by using solutions of heavy metal salts (*Table 4.1*) which, upon drying, form a glassy cast around the objects. It is assumed that the metal salt solution is able to occupy hydrated regions around and within the specimen, that is, the stain is able to penetrate into cavities. Ideally, the stain should not bind to the specimen, hence the name negative staining. In projection, the specimen is then represented by stain-excluding areas, that is, it remains "electron translucent".

Negative staining procedures may complement (or even replace) more elaborate techniques. Surface (S) layers on bacterial cells are mostly prepared by freeze fracturing of whole cells (cf. Sleytr and Messner, 1983 see also Section 4.5 and *Figure 4.7a*). As an alternative, the surface layers can be isolated by extraction of whole cells with 2% (w/v) sodium dodecyl sulphate (SDS) at 60°C and harvested by differential centrifugation. The extract contains large arrays of crystalline layers and may be prepared by negative staining.

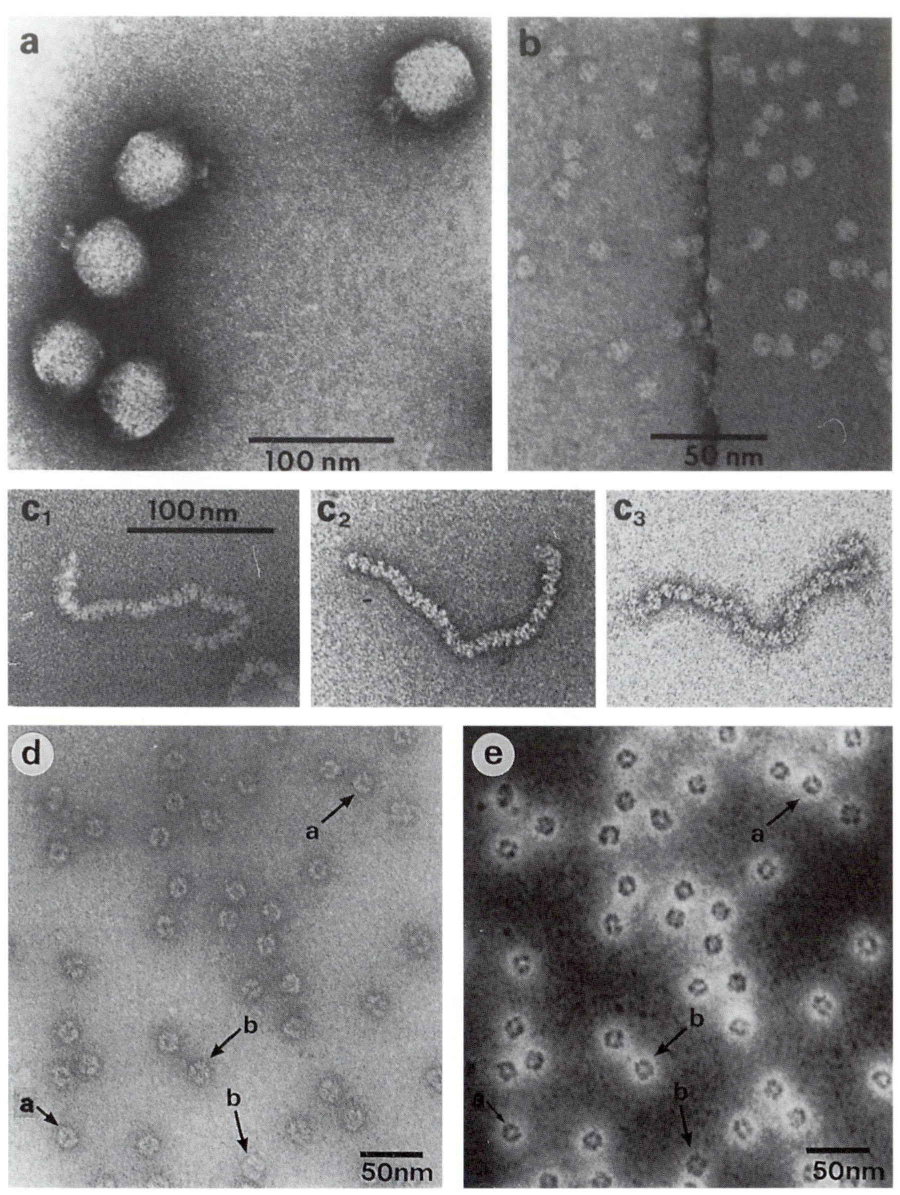

Figure 4.1. Negatively stained phages and isolated proteins. (a) Bacteriophage P 22 stained with 4% w/v uranyl acetate. (b) Hydroxylamine oxidoreductase from *Nitrosomonas europaea* stained with 4% w/v uranyl acetate; left half of the image: the molecules are supported by one carbon film (shallow stain); right half: the molecules are located between two carbon films (sandwich). See also Hoppert *et al.* (1995). (c) recA protein–pdT–DNA complexes (G. Möller, M. Hoppert and C. Urbanke, unpublished) stained with 4% w/v uranyl acetate in various stain depths: c_1 deep; c_2 medium; c_3 shallow stain. (d), (e) Large enzyme molecules (~ 790 kDa; F_{420}-reducing hydrogenase from *Methanobacterium thermoautotrophicum*), two different projection forms (a, b) are marked by arrows. Stained with 4% w/v uranyl acetate. In (e) a dark-field image, formed in an EFTEM (see Section 4.8) is used to enhance (and reverse) contrast. From Braks *et al.* (1994).

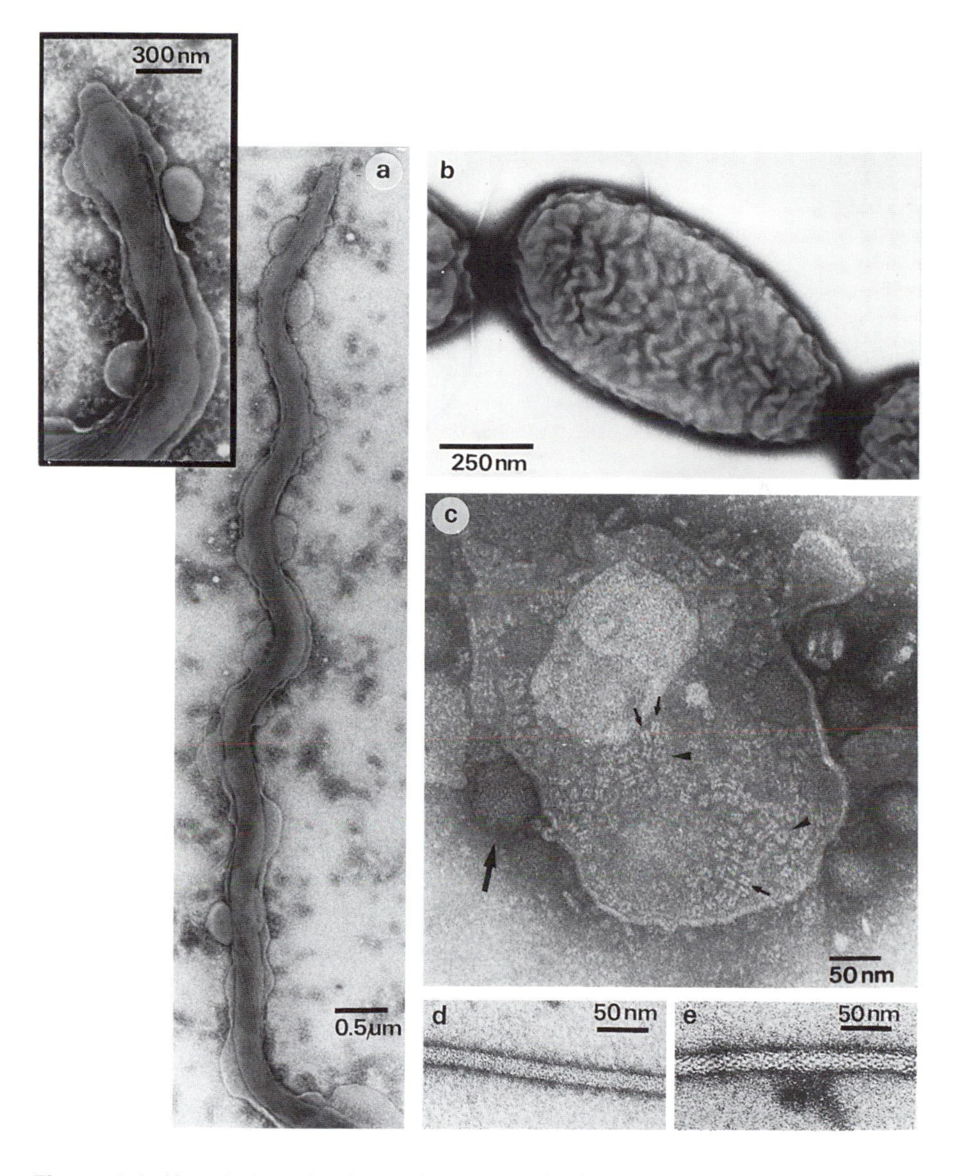

Figure 4.2. Negatively stained organisms and subcellular complexes. (a) Spirochaete *Borrelia burgdorferi*. Sample stained with phosphotungstic acid (3% w/v, pH 7.0). Inset: cell pole; the insertion sites of flagella in the cytoplasmic membrane are visible. (b) Bacterial cell depicting a heavy negative stain effect. Dried stain forms a dark rim surrounding the cell (see schematic drawing in *Figure 4.3b* for explanation) and is also entrapped in grooves on the wrinkled surface. Stained with phosphotungstic acid (3% w/v, pH 7.0). Original micrograph: F. Mayer. (c) Thylakoid vesicle from the cyanobacterium *Synechococcus leopoliensis*. "Rod" particles (small arrows) and "core" particles (arrowheads) of phycobilisomes as well as glycogen granula (large arrows) are visible. Sample stained with phosphotungstic acid (3% w/v, pH 7.0). (d), (e) Fine structure of flagellar filaments of Gram-negative hydrogen bacteria (stained with 2% w/v uranyl acetate). Original micrograph: F. Mayer: Aragno *et al.* (1977).

Table 4.1. Selected heavy metal solutions suitable for negative staining

Stain (reference)	Solutions and selected properties	Suitable for the visualization of
Uranyl acetate (Van Bruggen *et al.*, 1960, 1962)	Aqueous solution (0.5–4% w/v),pH (~ 4.2) adjustable to ≥ 2 using HCl or ≤ 4.8 using 0.1 M KOH (titrate carefully, above pH 4.8 insoluble $UO_2(OH)_2$ is formed); by adding EDTA it is possible to adjust the pH up to 7.4 with a concomitant increase in grain size; store in the dark	Macromolecules, viruses, appendages of bacterial cells (*Figure 3.1a–c*). Note: damage due to acidic pH may be possible (e.g. loss of membrane-bound F1-ATPase)
Uranyl formate (Lebermann, 1965)	Aqueous solution (~ 0.75% w/v), similar to uranyl acetate but smaller grain size, pH 3.5 without titration; pH can be titrated to 4.5–5.2 using ammonia; very unstable in solution (always prepare fresh); highest density uranyl salt	
Uranyl oxalate (Mellema *et al.*, 1967)	Uranyl acetate (0.5% w/v) in 12 mM oxalic acid, pH 5–7; pH adjustable with ammonia; stable at pH 6 for up to 2 days and at pH 7 for a few hours (at 4°C), solutions are light sensitive (store in the dark)	
Uranyl sulphate (Estis *et al.*, 1981)	Aqueous solution (2% w/v), pH 3.6; pH adjustable to 4.5 with dilute ammonia; titration to pH ≥ 5 causes precipitation; stable for weeks when stored in the dark at room temperature, characterized by a high radiation resistance (!); second highest density uranyl salt	
Phosphotungstic acid or corresponding Na and K salts (Brenner and Horne, 1959)	Aqueous solution (0.5–3% w/v) useful pH range 4–9, adjustable with 1 M KOH or NaOH; for the lower pH range use free acid; solutions are stable; tendency to positively stain an object (especially at lower pH)	Macromolecules[a], viruses, subcellular structures, membrane proteins, good recognition of refractile bacterial inclusion bodies (R bodies), bacterial cells (*Figure 3.1d*)
Methylamine tungstate (Fabergé and Oliver, 1974)	Aqueous solution (typically 2% w/v), useful pH range 3–10, can be used in connection with buffers	Macromolecules, viruses, appendages of bacterial cells, subcellular structures, whole bacterial cells

(Continued)

Ammonium molybdate (Muscatello and Horne, 1968)	Aqueous solution (2–10 % w/v) pH range from 5.2–7.5, titrate with NaOH/KOH or ammonia	As above[a] but particularly suitable for cell membranes, membrane-bound systems, protoplasts and other osmotically labile samples since the tonicity can be adjusted over a large range
Sodium silicotungstate	Aqueous solution (1–4% w/v) adjustable with 1 M NaOH[b]	Viruses
Mixtures of carbohydrates and negative stains (Harris and Horne, 1994)	Aqueous solution of 4% (w/v) uranyl acetate or 5% (w/v) ammonium molybdate and 1% (w/v) trehalose, ammonium molybdate-trehalose at pH 6.9 with NaOH; improved spreading behaviour (for details see reference)	Large macromolecules; there may also be possible advantages with bacterial cells, appendages of bacterial cells, etc.
Salts from light elements (Massover and Marsh, 1997)	Aqueous solutions of 2% (w/v) sodium tetraborate (pH 9.6), potassium aluminium sulphate (pH 3.4) or ammonium borate (pH 8.6)	Alternative to heavy metal stains (low electron beam doses required)

[a]When staining macromolecules, this stain may result in a lower contrast.
[b]Only available as silicotungstic acid. For best results, when preparing the corresponding salt, the following procedure is recommended: Solutions of silicotungstic acid and sodium hydroxide are combined in equimolar amounts. In practice, the NaOH solution is added dropwise and the pH monitored. The final pH should be 7. Now the solution is rapidly injected into ethanol at –20°C and left overnight. The white precipitate containing solution is filtered, the white deposit appearing on the filter subsequently washed with cold (–20°C) ethanol, dried into a cake (overnight) and finally ground to a fine powder. This powder can be stored and solutions are made up fresh as required at 1–4% (w/v) in double-distilled water.

Because of the ease with which negative staining may be performed, it is not only a powerful preparative tool for structural analysis but also for rapid screening/quality monitoring (e.g. checking the homogeneity of cell and protein preparations, or the contamination of cell cultures by viruses; see below).

Several variations of the negative staining technique (originally from Farrant, 1954) have been developed (Valentine *et al.*, 1968; Haschemeyer and Myers, 1972). The protocols in *Tables 4.2* and *4.3* describe two of the more successful modifications.

Positioning a specimen between two carbon films (see *Figure 4.1b*; “sandwich” technique) results in smaller variations in stain thickness around the specimen since the specimen is slightly squeezed between the carbon layers. The sandwich technique is used, if a more uniform stain distribution throughout the entire structure is desired. Variation in stain depth may provide additional information on the specimen structure. *Figure 4.1c* shows molecules in various stain depths. A “sandwich” preparation is achieved by using a rectangular piece of carbon-coated mica (measuring e.g. 3 mm × 6 mm) and turning the grid in such a way when picking it up from the staining solution (step 6) that the film folds back on to itself, thereby trapping the staining solution and the

Table 4.2. Negative staining protocol 1 (see *Figure 4.3*)

1. Pipette 50–100 μl droplets of sample, [fixing], washing and staining solutions on to a clean Parafilm™ surface or place solutions in small wells (e.g. microcentrifuge tube lids). The volume can be varied according to the amount of material available. However, it is important that whatever volume one uses, a pronounced convex meniscus is visible. As far as the concentration is concerned, trial-and-error is the best way to reach optimal results. As a guideline, isolated protein molecules with a molecular mass of 500 kDa should be used at a concentration between 70 and 100 μg ml^{-1} (the lower the molecular mass, the lower the concentration required) and viruses at about 10^{10}–10^{12} particles ml^{-1}; with bacteria an aliquot from the early log phase is adequate
2. Using sharp, pointed scissors cut an approximately 3 mm × 3 mm square from the carbon-coated mica. Holding the mica at one end with forceps, introduce it into the sample suspension at an angle of 30–45°. The carbon film will start to float off but is held in place by the forceps. During this step, particles are being adsorbed on to the carbon film. The adsorption time (usually in the range 10–60 sec) and the sample concentration determine the final particle density on the film
3. Remove the mica slowly, allowing the carbon to fall back into its original position. Drain off *excess* of liquid only, using filter paper. Do not blot dry!
4. Using the same procedure (step 2), introduce the mica into a fixing solution (0.1–0.5% v/v buffered glutardialdehyde). Drain off *excess* of liquid.
   N.B. Fixation may be omitted with most objects. The uranyl salt used in step 6 acts as a fixative
5. Using the same procedure, introduce the mica into a drop of distilled water or a suitable buffer in order to obtain a cleaner background. Drain off *excess* of liquid
6. Completely float the carbon off the mica and on to the staining solution (e.g. 4% (w/v) uranyl acetate, see *Table 3.3*) for 30 sec. Using forceps, pick up a grid and touch it to the surface of the carbon film. When the film attaches, lift up the grid and turn it around before blotting on filter paper. Complete removal of the staining solution results in a shallowly stained specimen. When some part of the staining solution remains on the carbon as a result of incomplete blotting, a deeply stained specimen is obtained
7. Allow the specimen to air dry

particles adsorbed between two layers of carbon film. While this effect may be desired under certain circumstances and when imaging macromolecules, it often tends to produce rather poor images, especially with large objects such as whole bacteria. With highly symmetric viral capsids, two-sided staining can give rise to Moiré effects and subsequently lead to a false impression of the arrangement of capsomers. Dr L. Stannard from the University of Cape Town demonstrates this effect on her ‘negative staining page’ at www.uct.ac.za/depts/mmi/stannard/negstain.html. For these reasons an unintentional folding back of the carbon film should be avoided.

In cases where the above protocol does not give satisfactory results, it is recommended that protocol 2 (*Table 4.3*) (which is particularly suitable for membranous samples and detergent-solubilized membrane proteins) is tried.

For the visualization of virus particles, traditionally, sodium and potassium phosphotungstate have most routinely been used. However, one should bear in mind that these stains may disrupt membranes. When preparing negatively stained specimens of enveloped viruses, it is therefore advisable to have an open mind and try other stains such as methylamine tungstate (see *Table 4.1*) or a solution of sodium silico-

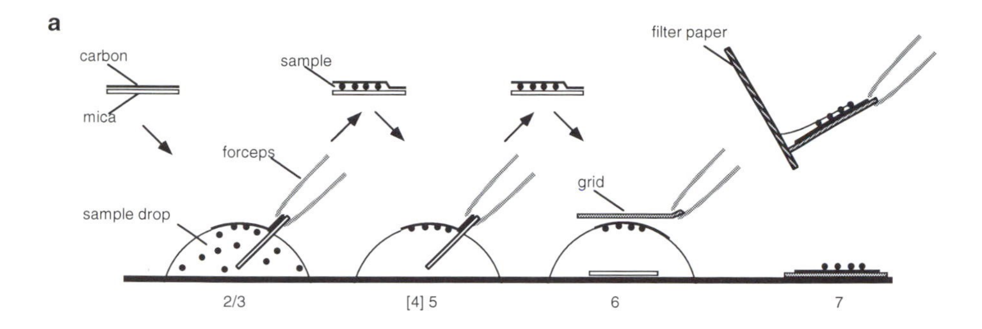

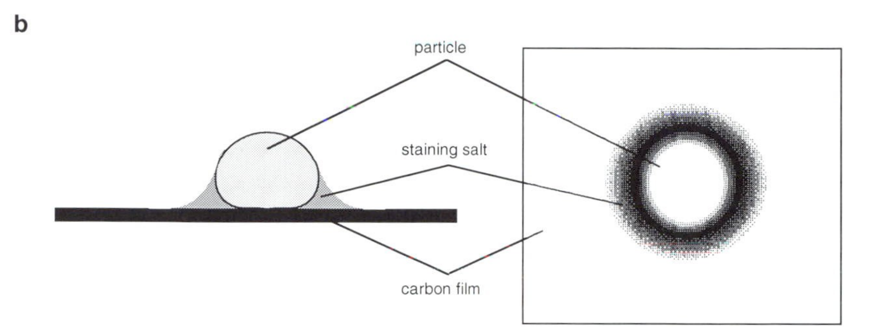

Figure 4.3. Negative staining procedure. (a) Schematic drawing referring to the procedure presented in *Table 4.2* (steps 2–7). (b) Sample particle attached to a carbon film and surrounded by dried stain. Left: side-on view; right: face-on view of the particle mimicking the corresponding projection as observed by TEM.

Table 4.3. Negative staining protocol 2

1. Pipette onto the clean surface of a strip of Parafilm™ 15–25 μl sample droplets (i) followed by two droplets of double-distilled water (ii,iii) and one droplet of, e.g. a 4% (w/v) aqueous solution of uranyl acetate (iv)
2. Place carbon-coated collodion grids (freshly glow discharged for 30 sec under a reduced atmospheric pressure of 0.1 Torr — this is an option), carbon side towards the sample, on to (i) for 20–50 sec and quickly blot to remove *excess* of liquid. This is the adsorption step
3. Place grid on to (ii) and (iii) for about 1 sec each (each followed by a blotting step, but without blotting dry). This is the washing step
4. Stain specimen by placing it on (iv) for 30 sec
5. Blot dry with the grid perpendicular to the blotting paper or use modifications according to protocol 1 (*Table 4.2*)
6. Allow specimen to air dry

tungstate at pH 7. It is important to remember that the success of the staining procedure, particularly with respect to enveloped viruses, is not only influenced by the stain employed but also by the surface properties of the support film. Uranyl acetate, for instance, is normally not the preferred stain for enveloped viruses, but using it in connection with a negatively glow-discharged support film (see Section 3.2.1) can deliver remarkable results. It may therefore become necessary to fine tune the protocol in order to meet the exact requirements of a particular virus. To this end, the protocols provided in *Tables 4.2* and *4.3* can be regarded as promising starting points for optimization.

Electron microscopy was used early on in the elucidation of the nature of viruses and also for rapid viral diagnosis (e.g. smallpox v. chicken pox; see also Almeida, 1980). Nowadays, in spite of all modern molecular biological and genetical tools, electron microscopy is still useful, for instance when dealing with emerging virus diseases, stool viruses, double infections, cell culture cytopathic effects or hepatitis B virus in blood. It allows in many cases a rapid identification directly in clinical specimens, and the test results can often be further serologically characterized by immunoelectron microscopy along the lines described in Section 4.7 or simply by adding a specific immune serum to the sample solution and observing the agglutination of virus particles.

It should also be mentioned at this point that negative staining is not the only way forward. For instance, the diagnosis of progressive multifocal leucoencephalopathy (PML) can be confirmed by electron microscopy of thin sections of oligodendrocyte nuclei where the papovavirus virions often occur in conspicuous crystalline arrays. Thin sectioning (see Section 4.6) has the advantage that the virus particles can be detected directly within the cell obviating the need for, for example, concentration or isolation. Whatever way forward one chooses, in order to guarantee a good diagnostic standard, it is advisable to regularly join external quality assessment schemes, such as those run by the Central Public Health Laboratory Service, Colindale (UK) or by the Robert Koch-Institut, Berlin (Germany).

For further up-to-date information on electron microscopy of viruses, the reader may wish to consult the Internet. A page worth looking into with many first-class site links, such as the one referred to above, is www.vet.ed.ac.uk/restrict/path1/virology.htm.

4.2 Visualization of nucleic acids

In most cases, isolated nucleic acids (double-stranded [ds] or single-stranded [ss] DNA or RNA molecules) have to be prepared with the aid of special spreading techniques before staining and visualization in the electron microscope (Spiess and Lurz, 1988; see *Figure 4.4b* and *Table 4.4*). Contrast is usually introduced by rotary shadowing of the specimen. Because of its small diameter, DNA protrudes only very little from the surface, necessitating a reduction in the shadowing angle to 5°–7°. A detailed description of a simple and reproducible mounting procedure for DNA on mica sheets is given in *Figure 4.5* and *Table 4.5*. Depending on how densely the nucleic acids are "decorated" with (and hence stabilized by) proteins, nucleic acid–protein complexes may even be prepared and stained using a negative staining protocol (see, e.g.

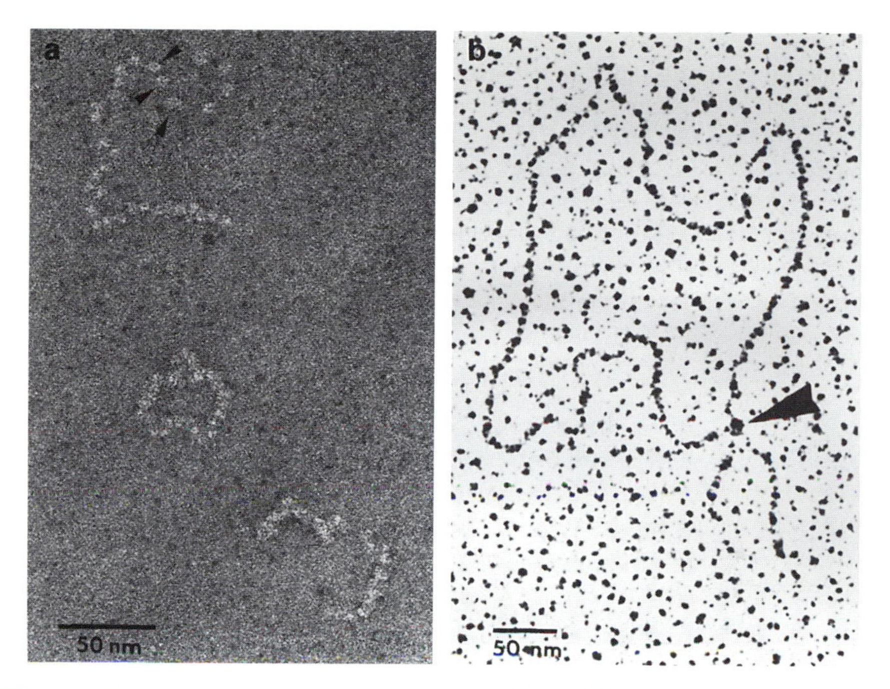

Figure 4.4. Visualization of nucleic acids and nucleic acid–protein complexes. (a) ss Binding protein (arrowheads)–pdT–DNA complexes, prepared by conventional negative staining (G. Möller, M. Hoppert and C. Urbanke, unpublished); the DNA is not visible. (b) dsDNA (linearized pBR 322) with one EcoRI-molecule bound (arrowhead); prepared by BAC spreading (see *Table 3.5*) and rotary shadowing with Pt–C (original micrograph: H. Gerberding). Note the loop formation of the DNA caused by the action of the restriction endonuclease.

Greipel *et al.*, 1987, *Figure 4.4a*). Modifications of this technique have been used to visualize ss nucleic acids (cf. Spiess and Lurz, 1988) and other filamentous molecules (Tyler and Branton, 1980). Note that highly structured nucleic acids (e.g. certain RNA molecules containing multiple stem-loops) can also be visualized completely "naked" using negative staining protocol 1 (*Table 4.2* and *Figure 4.3*). However, one should be aware that in some cases the molecules may stain positively.

4.3 Sample preparation by rapid freezing

It is well known that, although negative staining salts and chemical fixatives act as fixing and stabilizing agents, macromolecules and whole cells suffer from distortion caused by dehydration. In order to preserve samples in their fully hydrated state, samples must be frozen at very high cooling rates and, from then onwards, be kept at temperatures below –150°C until further processing occurs.

Table 4.4. Preparation methods for different nucleic acids

Method	Preparation	Suitable for visualization of
Conventional negative staining	DNA-containing aqueous (buffered) solution prepared as described above (Greipel *et al.*, 1987; Mayer and Friedrich, 1986)	Nucleic acid–protein complexes (ds DNA barely visible, ss DNA invisible; *Figure 4.1a*)
Droplet diffusion technique	Sample is picked up with a carbon (or collodion; see *Table 3.2*) -coated grid from a droplet of ammonium acetate buffer containing formamide and cytochrome *c* (molecular mass 12 kDa) or benzyldimethyl-alkylammonium chloride (BAC; molecular mass ~350 Da) and DNA (Lang and Mitani, 1970; Vollenweider *et al.*, 1975)	ds DNA, nucleic acid–protein complexes when using BAC; *Figure 4.1b*
Support film pretreatment	Sample is applied to a carbon-coated grid pretreated with Alcian blue 8GX [purify before use according to Scott (1972)], which enhances the binding of nucleic acids to the support film (Labhart and Koller, 1981)	Nucleic acid–protein complexes, nucleoids
Carbonate spreading	A droplet of carbonate buffer containing formamide, cytochrome *c* and the DNA sample is spread on a water surface and picked up with a coated (see above) grid (Westmoreland *et al.*, 1969; Rochaix and Malnoe, 1978)	ss DNA, ds DNA, RNA–DNA hybrids, replication intermediates
Mounting on to mica	Specimen is adsorbed onto the surface of a mica sheet from a droplet containing the nucleic acid in Tris–acetate buffer (see *Figure 4.5*) (Spiess and Lurz, 1988)	ds DNA (< 30 kb), ss DNA and ss RNA, protein–DNA complexes

Table 4.5. Outline method for spreading nucleic acids on mica

1. Place a drop of the nucleic acid-containing solution (0.1–1 $\mu g\ ml^{-1}$ DNA or RNA in 5 mM Tris–acetate buffer, pH 7.5, containing 5 mM magnesium acetate) on the surface of the Parafilm
2. Cover the drop, for 1 min, with a sheet of freshly cleaved mica. Remove excess of solution by immersion in water
3. Incubate the sheet, for 1 min, in 1% (w/v) uranyl acetate solution and wash by transfer to water (2–3 times for 1 min)
4. Dry the mica sheet (face down on a filter paper)
5. Perform rotary shadowing of the sample with Pt–C and then, in a second step, carbon coating
6. Float off the carbon film on a water surface and then pick up the film with a bare copper grid

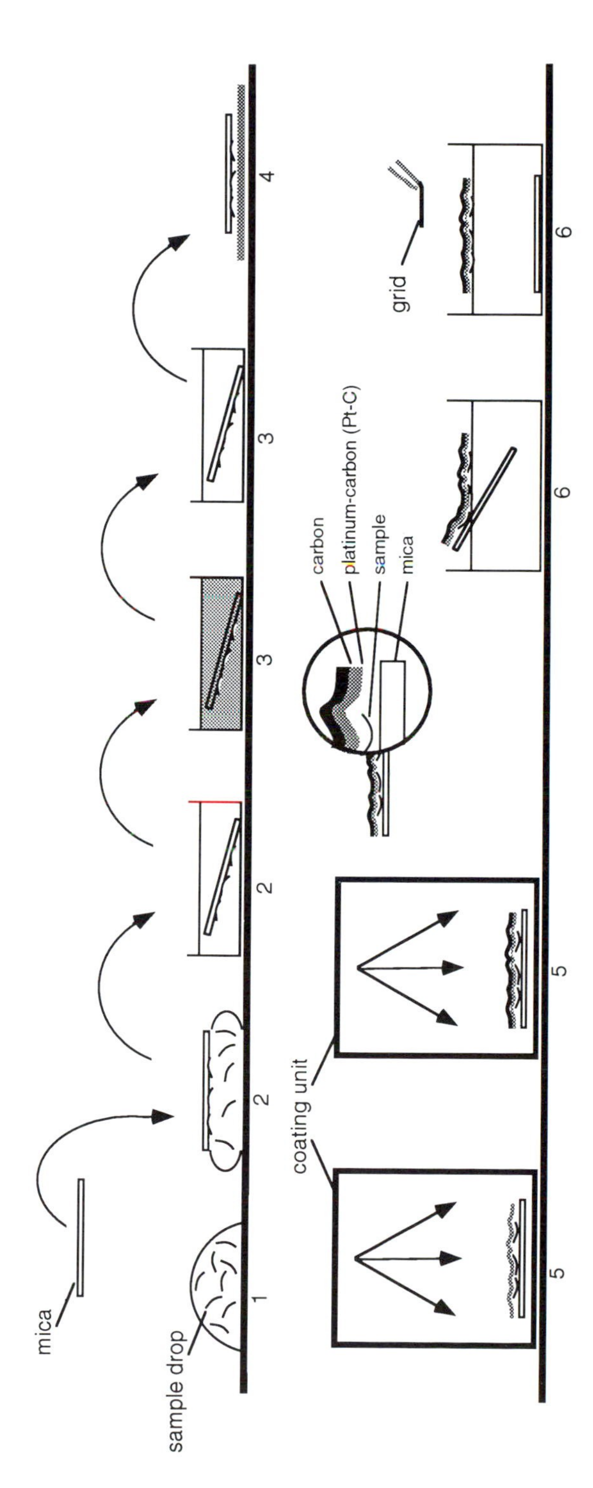

Figure 4.5. Mounting of nucleic acids on to mica. Schematic drawing refers to the procedure presented in *Table 4.5* (steps 1–6).

Slow cooling rates (below 1 K sec^{-1}) result in the formation of large ice crystals outside cells, which successively dehydrate the cells during the crystal growth. This results in increasing concentration of solutes inside cells and, of course, severe structural change (shrinkage). Nevertheless, the high solute concentration prevents formation of large intracellular ice crystals and leads to high cell survival rates.

Cooling rates of between 1 and 1000 K sec^{-1} are too fast for cellular dehydration by formation of extracellular crystals, that is, large ice crystals are also formed inside cells, leading to, for example, membrane disruption, resulting in a reduced survival rate. Cooling rates faster than 1000 K sec^{-1} give rise to smaller ice crystals. Above 10 000 K sec^{-1}, cells are structurally well preserved and show high survival rates. The object is embedded in microcrystalline ice or, under ideal conditions, amorphous ice (i.e. the sample is surrounded by a water network that is similar to liquid water, immobilized in a "frozen-hydrated" state also termed as "vitrified water") which does not damage its structure, the prerequisite for electron microscopy of these samples. Since heat transfer rates in biological specimens are low, the cooling rates are primarily influenced by sample size which has to be as small as possible. Thus, high cooling rates are relatively easy to achieve with suspensions or thin layers of biological macromolecules, viruses or single microbial cells, but difficult in complex tissue.

Cryofixation without pretreating the sample with cryoprotectant and without use of specialized apparatus that provides ultra-high cooling rates is only achievable with single cells. Cryoprotectants overcome the problem of ice crystal formation in large specimens or when rapid-freezing devices are not available (Skaer, 1982). When penetrating agents (glycerol, dimethyl sulphoxide or ethylene glycol) are used as cryoprotectants, artefacts such as irreversible plasmolysis, swelling, phase separation and membrane rearrangements may occur (Sleytr and Robards, 1981). Electrolytes and other solutes are redistributed. Non-penetrating agents (polyvinylpyrrolidone, hydroxyethylstarch or dextran) may induce shrinking artefacts by dehydration (Allen and Wheatherbee, 1979).

Specimens which have been fixed via cryo-immersion can be used in techniques requiring, for example, frozen-hydrated, freeze-sectioned, freeze-dried and freeze-fractured specimens (*Figures 4.6* and *4.7*). For comprehensive treatises on this subject, the reader is referred to publications by Dubochet *et al.* (1982, 1988), Robards and Sleytr (1985), Sitte (1984) and Stuart (1991).

4.3.1 *Pretreatment of subcellular components*

Rapid freezing of protein molecules, virus and subcellular compartments lead to excellent structural preservation at the molecular level (Plattner and Bachmann, 1982). Rapid cooling rates can only be obtained when thin films of particle suspensions (< 200 nm) and appropriate cryogens which allow rapid cooling rates are used. Only liquefied gases which

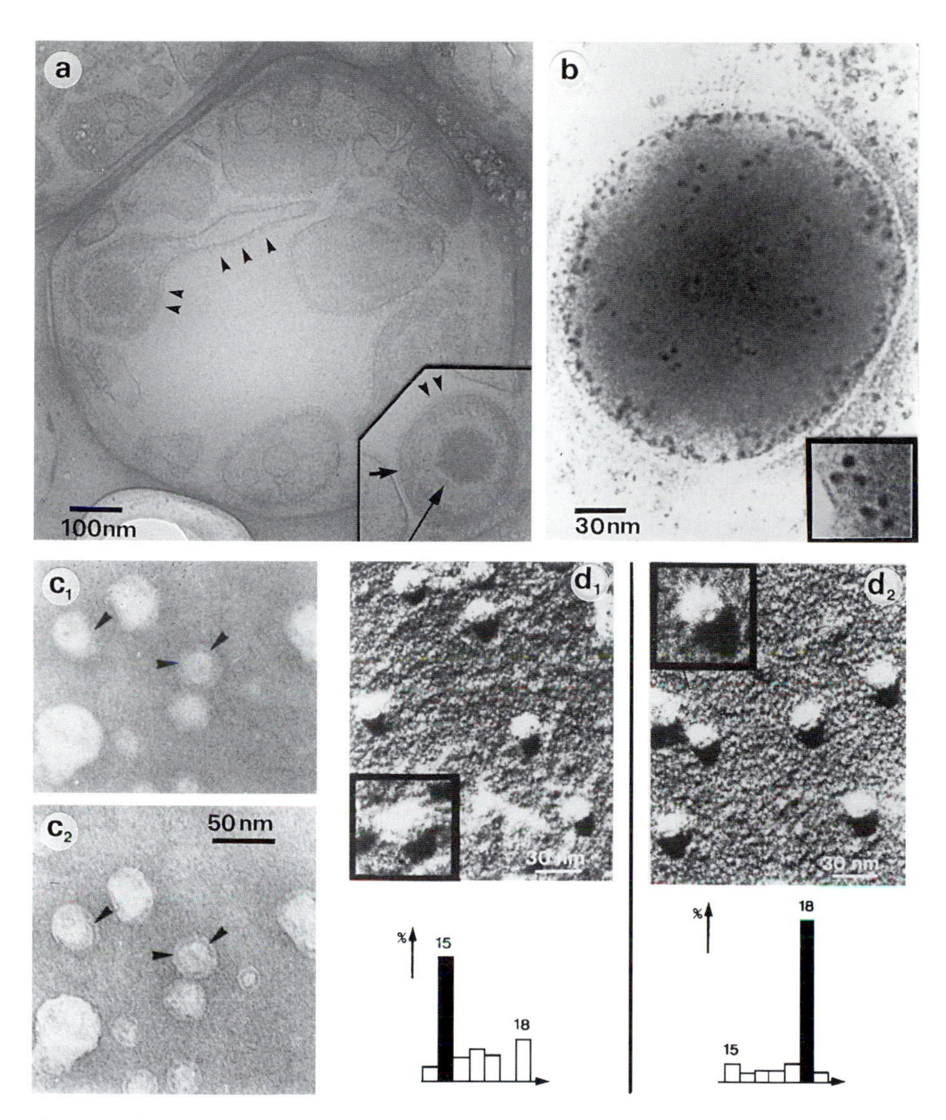

Figure 4.6. Application examples for freezing procedures. (a) Frozen hydrated Herpes simplex virus particles on holey carbon. The long arrow points at the capsid (inset: note the regular array of capsid proteins) and the short arrow at the envelope in which individual glycoproteins (arrowheads) are discernible (original micrograph: S. Stoylova). (b) Frozen hydrated enzyme molecules (citrate lyase). A thin film of vitrified ice enclosing the molecules spans over a hole in a holey plastic film (see Section 3.1). Inset: enlarged section showing one form ("star") of the molecule (Preparation by F. Mayer and LEO application laboratory, Oberkochen). (c) Proteoliposomes prepared by plunge freezing in liquid propane and further processed by cryosectioning and post-stained 0.3% (w/v) uranyl acetate in methylcellulose (see Sections 4.6.3 and 4.6.4). Electron micrograph taken in focus (c_1) and slightly underfocused (c_2), demonstrating the effect of phase contrast. Arrowheads: membrane contours (Bodenstein 1991). (d) Molecules of D-ribulose-1,5-bisphosphate carboxylase/oxygenase (RuBisCO) from *Ralstonia eutropha* (550 kDa) after freeze drying and unidirectional shadowing with W–Ta at an elevation angle of 45°. A slight variation in size between two different functional states is reflected by statistical analysis of the shadow width as indicated in the figures: d_1 15 nm, d_2 18 nm. From Holzenburg and Mayer (1989).

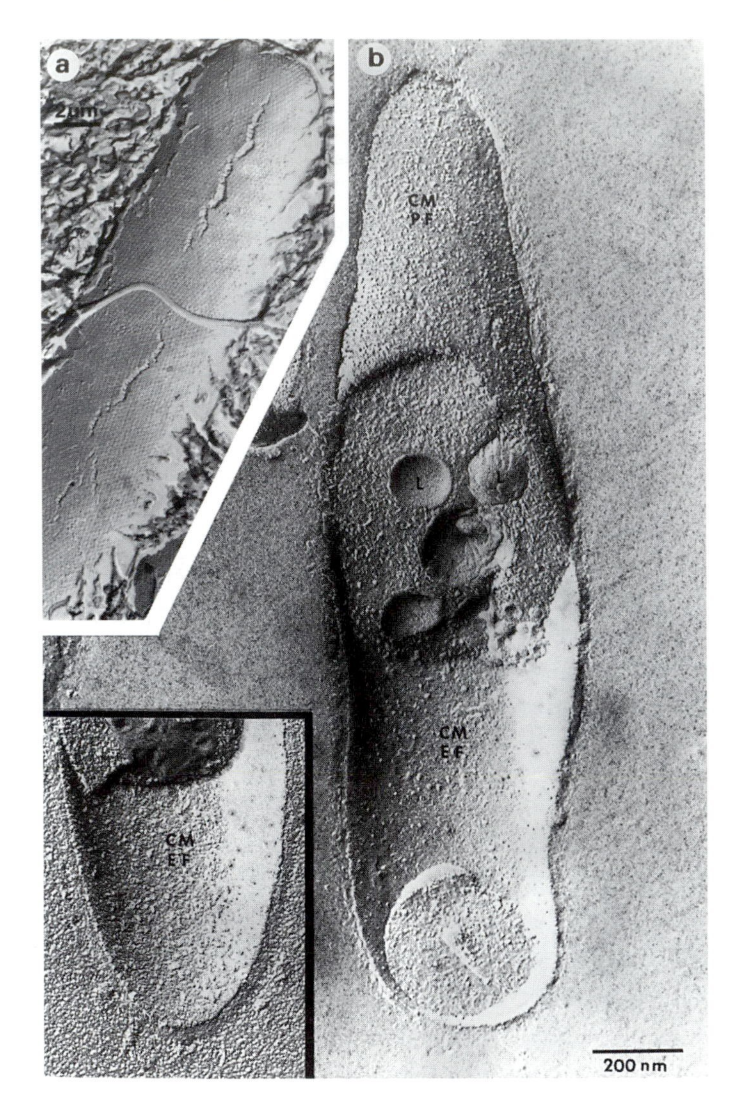

Figure 4.7. Freeze-fractured and freeze-etched specimens. (a) Freeze etched cell of *Aquifex pyrophilus*, exhibiting a surface layer (hexagonal symmetry) and a flagellum lying on top (original micrograph: R. Rachel; see also Huber *et al.*, 1992). (b) Freeze-fractured and freeze-etched (inset) preparation of *Streptomyces lividans* (see also Packter and Olukoshi, 1995). The cytoplasmic membrane (CM: EF — exoplasmic face, PF — plasmic face) as well as distinct lipid inclusion bodies (L) residing within the cytoplasm are clearly revealed. Note that on etching (inset), the membrane boundaries and particles on the membrane surface become more clearly visible as they are freed from surrounding ice. Shadowing source is at a 5 o'clock position (original micrographs: D. Ashworth).

do not show "film boiling" (Leidenfrost phenomenon, causing a reduced cooling rate because of the production of a gas layer around the specimen) can be employed as cryogens.

There are several alternatives for producing very thin films of particle suspensions. Grids with hydrophilized holey films or bare hydrophilized grids (G 400–G 600 mesh) are used (Adrian *et al.*, 1984, see *Table 4.6*).

Table 4.6. Performance of thin specimen films

Specimen film on a thin support film Place a small droplet (3–10 μl) of the particle suspension on the grid surface and suck off excess of liquid. Only a thin film of liquid should remain on the grid surface. The grid is immediately processed by plunge freezing (see below)
Bare grid method Apply a small droplet of suspension on a bare hydrophilic grid. Suck off excess of liquid, so that the grid squares are filled by thin aqueous films. Further thinning of the films is achieved via evaporation of the liquid for several seconds. The procedure may be checked using a stereomicroscope: when about 50% of the squares have lost their films, the grid is immediately plunged into the cryogen (see below) *Comment:* Vitrified films may be directly visualized in electron microscopes equipped with cryostages. Cryostages are available for most modern electron microscopes, but their operation needs some experience. Alternatively, they can be visualized, after freeze drying and metal shadowing (see below), by conventional TEM

4.3.2 Pretreatment of cells

Because of their small size, their relatively low water content and their relatively high mechanical stability (as compared to, e.g. animal tissue), treatment of prokaryotes prior to cryofixation need not be elaborate to avoid artefacts. If ultra-rapid freezing methods are applied, washed cell suspensions may be subjected to cryofixation without any further treatment. Chemical fixation and cryoprotection may be omitted, that is, chemically induced artefacts can be avoided. This factor is important for immunocytochemical localization of some aldehyde-sensitive antigens. If cryofixation without cryoprotection and/or chemical fixation does not lead to satisfying results or is not applicable, proceed as outlined in *Table 4.7.*

If centrifugation steps prior to fixation and/or freezing influence cell structure (e.g. loss of cellular appendages) or interactions between cells, an alternative method for cell concentration can be applied (Hohenberg *et al.*, 1994; Rieger *et al.*, 1997). Very high cell densities of several species of hyperthermophilic archaea were achieved by concentration of the cells in cellulose capillary tubes. The tubes are filled with cell suspension and incubated in growth medium. The capillaries allow the growth medium

Table 4.7. Cryoprotection of cells

1a.	For cells with a high water content (algae, protozoa), chemical fixation is necessary prior to glycerol infiltration. Incubate cells with buffered glutardialdehyde solution (1% w/v–4% w/v) for 90 min at 0°C. Usually, a 1% w/v solution provides sufficient fixation. If the cells are to be subjected to immunolocalization procedures, incubate with a mixture of 0.2% w/v paraformaldehyde and 0.3% glutardialdehyde for 90 min at 0°C (see below)
1b.	For prokaryotes and yeasts, attempt to culture the organisms in the presence of 20%–30% glycerol in the growth medium. Alternatively, resuspend cells in 20–30% buffered glycerol solution (adjust pH and ionic strength to the respective values of the growth medium) and incubate for up to 1 h. Cells should not be plasmolysed after the procedure
2.	Centrifuge the cells. When no cryoprotectant has been used, proceed with a densely packed cell pellet

Table 4.8. Cryoprotection of cells according to Tokuyasu (1986)

1. Incubate a washed cell pellet in a mixture of 0.2% (w/v) paraformaldehyde and 0.3% glutardialdehyde in an appropriate buffer (if there are no special requirements needed, 50 mM potassium phosphate is recommended) for 90 min at 0°C (see below)
2. Centrifuge the cells and resuspend in a small volume of a liquefied 10% (w/v) gelatin solution (in buffer)
3. Pour the solution on to a flat, chilled surface
4. Cut the solidified gelatin into small prism-shaped blocks (this shape will facilitate the trimming procedure prior to sectioning; see below)
5. Incubate the small blocks in buffer for 1 h
6. Incubate in a mixture of 0.2% (w/v) paraformaldehyde and 0.3% (v/v) glutardialdehyde for 8 h or overnight and then incubate in a solution of 1.6 M sucrose and 25% (w/v) polyvinylpyrrolidone for 8 h or overnight
7. Attach the sample to an appropriate specimen holder (depends on the equipment used, see below) and then freeze in liquid nitrogen or plunge freeze (see below)

to diffuse freely throughout the capillaries, and the organisms are retained and may be cultivated to higher cell densities than in regular culture. The capillary tubes are then cut into approximately 3 mm segments and submitted to rapid freezing. Since the capillary tubes are too large to permit rapid sample cooling rates by normal means, they have to be subjected to high-pressure freezing requiring specialized devices or be subjected to chemical fixation prior to plunge freezing (see Rieger *et al.*, 1997).

A cryoprotection procedure allowing the use of low cooling rates (see below, "freezing in liquid nitrogen") and requiring chemical prefixation of cells is mainly used for samples for cryosectioning and immunolocalization (Tokuyasu, 1986; *Table 4.8*).

4.3.3 Freezing methods

Freezing in liquid nitrogen. This method is used mainly for specimens to be subjected to cryosectioning. Only samples that have been completely infiltrated with cryoprotectant (sucrose/polyvinylpyrrolidone; see *Table 4.8*) may be frozen directly in liquid nitrogen. Since cooling rates are low, unprotected samples would be destroyed by ice crystal formation. The method does not require any specialized equipment.

After the sample has been attached to the appropriate specimen holder (depends on the cryo-ultramicrotome used, see below), the sample is placed in a cryo cap (or screw-capped microcentrifuge tubes as depicted in *Figure 4.8*) and covered with liquid nitrogen. The sample is stored in liquid nitrogen until use.

Plunge freezing. Plunge-freezing devices for small specimens with entry velocities of around 1 m sec^{-1} are easy to construct by a workshop. In a simple plunge-freezing apparatus, a cryogen (Freon 12, Freon 22, ethane or propane) is held in an aluminium cylinder surrounded by liquid nitrogen. Care has to be taken that the cryogen remains liquid (propane: m.p. –189°C, b.p. –42°C; ethane: m.p. –172°C, b.p. –89°C).

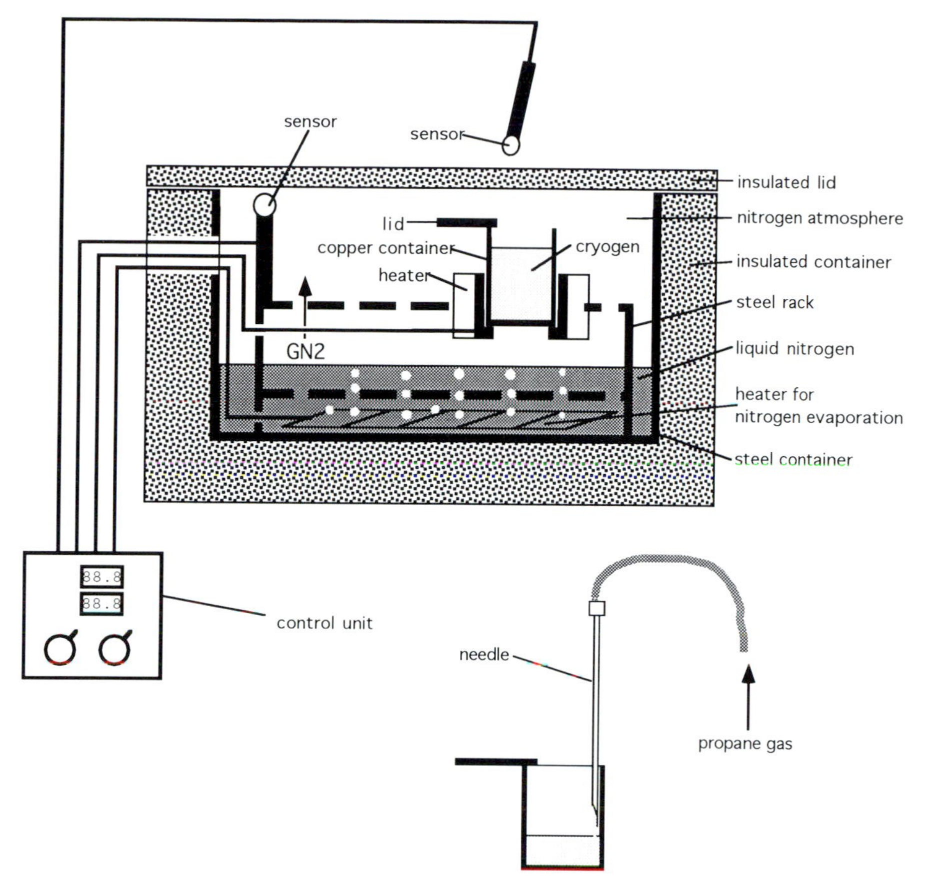

Figure 4.8. Apparatus for rapid freezing of samples in liquid cryogen (see Section 4.3.3).

Stirring of the cryogen during plunge freezing facilitates heat transfer. The essential design of a more advanced plunge-freezing device in liquid propane or ethane is given in *Figure 4.8* (the plunge-freezing procedure is outlined in *Table 4.9*). The metal cylinder is attached to a steel rack that is cooled by liquid nitrogen. The cylinder itself is not immersed in nitrogen, to allow electrical counter-heating for exact temperature adjustment. The outer insulating container traps an atmosphere of nitrogen gas (produced from liquid nitrogen by a small heated wire submerged in nitrogen) that surrounds the cylinder and all tools to be precooled in the chamber and prevents entry of moisture, that is, ice contamination is avoided. Temperature control by sensor circuits increases the reproducibility of plunge fixation. A built-in sensor reports the ambient temperature of the chamber. A second, removable, sensor is used to check the cryogen temperature. Propane or ethane gas is liquefied as depicted in *Figure 4.8*. Gas flow must be initially adjusted to a low rate. The gas is ducted via a long injection needle into the nitrogen-chilled

Table 4.9. Plunge freezing procedure

1. Before freezing, prepare the plunge-freezing device and the necessary instruments. Liquefy the cryogen (e.g. at −180°C for propane). Pre-chill all tools that will come into direct contact with the frozen sample (forceps, cryotransfer tools, etc.) in liquid nitrogen. Ensure that the chilled parts of the instruments are free of ice crystals. This is achieved by keeping the instruments in the layer of nitrogen gas settled above the liquid nitrogen. Transfer times of the instruments in air have to be reduced to a minimum. Contact of the frozen specimen with air has to be avoided throughout the complete procedure
2. Apply a small drop of the concentrated cell suspension or gelatin embedded block to a bare specimen grid, a small (3 mm^2) piece of filter paper, a small plunger or a wire loop (it depends on the subsequent preparation steps) and mount it on to the plunger of the plunge fixation device (see *Figure 4.9*)
3. Inject the sample into liquid propane
4. Withdraw the frozen sample from the propane vessel. If a drop of liquid propane covers the sample, remove it with a small piece of chilled filter paper that is kept in the cryochamber. Residual propane may cause problems during the subsequent preparation procedures (especially during preparation of frozen-hydrated samples). All operations with the sample have to be done gently, because the frozen sample and the support are very brittle
5. Transfer the sample according to the processing steps given below. Samples that are to be processed by freeze substitution or cryosectioning may be kept in liquid nitrogen for several days.

(not counter-heated) cylinder. In order to avoid contamination by moisture, the cylinder has to be closed by a lid. Be aware that operation with propane does require extensive safety measures (see Appendix D).

A mechanical spring-driven device for injecting specimens into liquid cryogens or slamming them onto metal mirrors (see below) is depicted in *Figure 4.9*. Various injection mechanisms, driven by electromagnets, compressed air or (most simply) gravity do exist.

Liquid cryogen residues on samples may interfere with subsequent preparation steps (e.g. cryo-electron microscopy of frozen-hydrated samples or cryosectioning). These problems are avoided by the use of subcooled nitrogen as cryogen. Though liquid nitrogen at its equilibrium boiling point is not a good coolant, with subcooled nitrogen, the “film boiling” does not occur. A device for preparation of subcooled nitrogen is described by Umrath (1974). A container of liquid nitrogen, built as a vacuum chamber, surrounds a copper container also filled with liquid nitrogen, but open to the atmosphere. On evacuation, the nitrogen in the surrounding vacuum chamber cools the liquid nitrogen in the copper container, which can, therefore, be maintained continuously at a temperature of around 59 K and used for plunge fixation.

Besides subcooled nitrogen, nitrogen slush (i.e. a mixture of solid and liquid nitrogen) is used as a cryogen. Its preparation is very simple, but because of the heterogeneity of the solid–liquid mixture, reproducibility of the cooling rates in this cryogen is low. The nitrogen container is placed in a vacuum chamber and evacuated under a vacuum provided by a one-stage rotary pump. Nitrogen solidifies at a temperature of 63 K and a pressure of 13.5 kPa. After readmission of air, the mixture of liquid and solid nitrogen remains stable for some minutes.

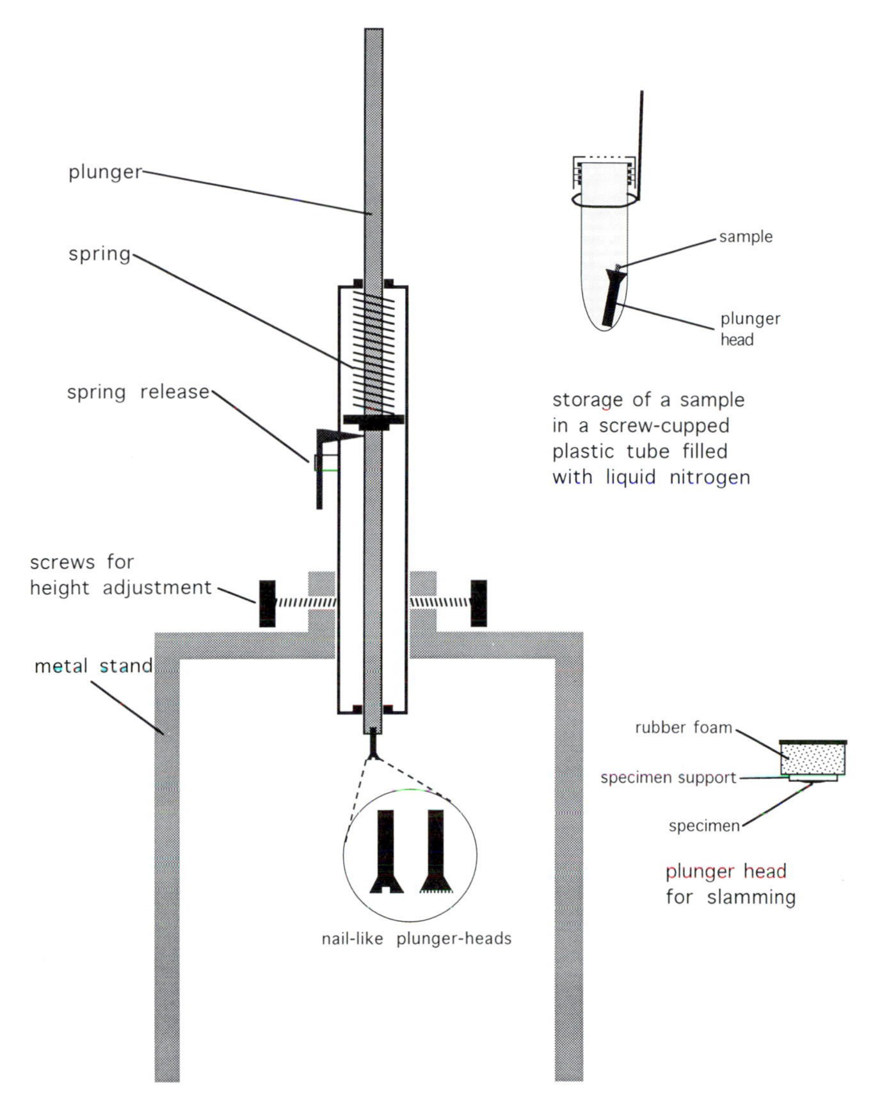

Figure 4.9. Plunge freezing. A mechanical spring-driven device for injecting specimens into liquid cryogens or slamming on to metal mirrors (see below). The nail-like plunger heads for sample application are preferably used for subsequent cryosectioning.

Sample application by spraying grids prior to plunge freezing. The suspension is sprayed from a nebulizer on to a grid with a hydrophilic support film immediately before freezing. This is normally done by spraying the grid during the fall just prior to plunging into the cryogen.

Spray freezing. The freezing rate is increased by minimizing the sample size. Therefore, rapidly frozen aerosols of small samples (e.g. viruses or bacterial cells) with a droplet size of 10–50 μm in diameter provide excellent results (see procedure in *Table 4.10*). Of course, this technique

Table 4.10. Spray freezing procedure (see *Figure 4.10*)

1. Liquefy propane in a copper (or brass) block at –180°C and cover the block opening with a lid to avoid contamination
2. Suspend a cell pellet in the same volume of an appropriate buffer solution (see above). Fill the airbrush with 200–300 μl of the sample
3. Cover the propane container with a plate, the opening of which is slightly smaller as the opening of the container (as indicated in *Figure 4.10*). The plate protects propane from contamination with ice formed during the spray process
4. Place the nozzle of the airbrush at a distance of 0.5–1 cm from the container and spray the sample in the container at short intervals. Raise the container temperature to –85°C
5. Remove the plate and evaporate the propane by connecting the container opening to a one-stage vacuum rotary pump
6. For freeze fracturing (see below), suspend in butylbenzene. The samples are then transferred to specimen holders and may be stored in liquid nitrogen, which solidifies the butylbenzene (i.e. entraps the sample). Further processing by freeze substitution is also possible

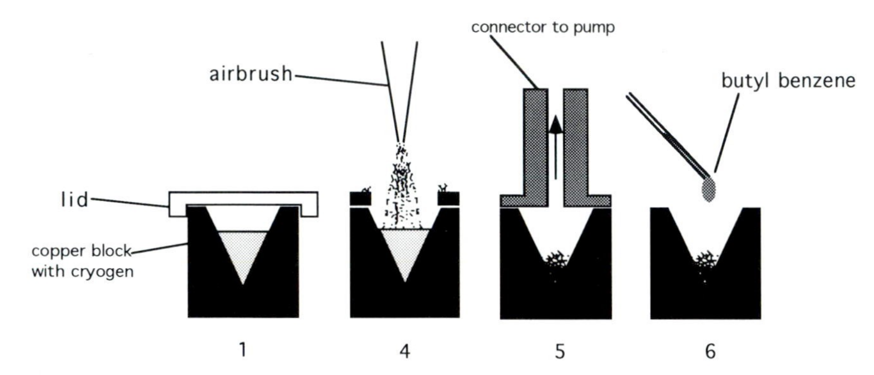

Figure 4.10. Spray-freezing procedure. The figures refer to preparation steps given in *Table 4.10.*

is restricted exclusively to small particles in suspension. Construction of spray-freezing devices is somewhat more complicated than that of simple plunge fixation devices and creates some difficulties in handling of the frozen sample (Bachmann and Schmitt, 1971; Plattner and Bachmann, 1982). Spray freezing turned out to be very helpful in capturing rapid dynamic cellular events at defined intervals after rapid mixing with a stimulating agent (Knoll *et al.*, 1991). A device specially designed for these experiments is offered by Balzers (model SFD 010).

Spray freezing is mainly used in connection with subsequent freeze fracturing and freeze etching. A commercially available artist's airbrush may be used as atomizer of the cell suspension.

Freezing on cold surfaces (cold block freezing, "slamming"). Heat transfer between a cooling agent and the sample may be increased by use of a solid (instead of a liquid as secondary "cryogen"). Heat-transfer rates of very cold metals (especially copper) are very high and they are, therefore, suitable for specimen freezing. The optimum temperature for rapid

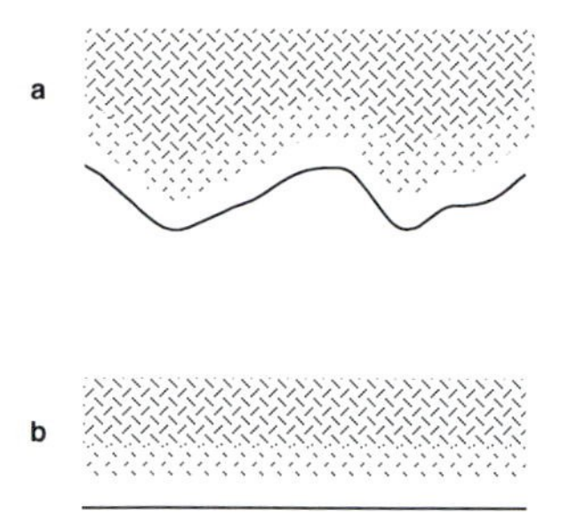

Figure 4.11. Distribution of ice crystals (small dashes) in samples prepared by plunge freezing (a) and cold block freezing (b).

cooling on a cold copper block is 19 K, that is, for the fastest cooling rates, liquid helium must be used instead of liquid nitrogen. The technique is primarily used for the cryofixation of tissue. However, the formation of an absolutely planar sample surface during preparation, as occurs during the technique, is advantageous for any sample type. Planar parallel layers extending 10–20 μm from the surface are vitrified, and deeper regions are damaged by large ice crystals and compression shock (*Figure 4.11*).

In principle, "slamming" is performed via a modified plunge fixation protocol. A simple version of the apparatus required is depicted in *Figure 4.12* (Van Harrefeld and Crowell, 1964; Van Harrefeld *et al.*, 1974). A highly polished, scratch-free metal block (metal mirror) is chilled with liquid nitrogen. An atmosphere of nitrogen gas keeps the mirror ice free. The plunger is driven by a spring (see *Figure 4.9*). The specimen has to be cushioned from the full force of impact by a piece of rubber foam attached to the metal holder. The specimen support may be a small piece of filter paper or freshly cleaved mica attached to the foam with double-sided adhesive tape. Care has to be taken that the metal

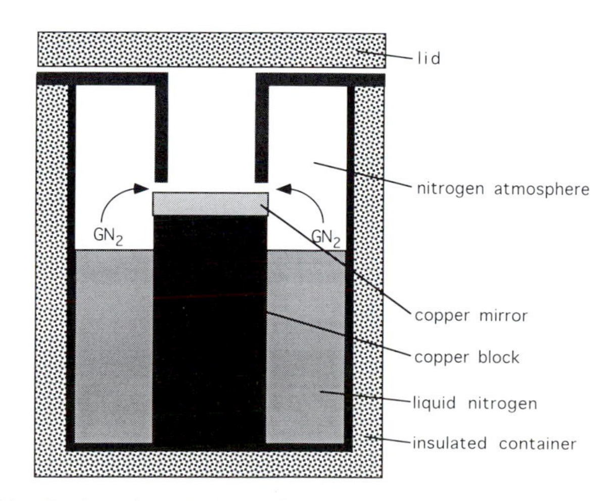

Figure 4.12. Metal mirror for cold block freezing or slamming (see Section 4.3.3).

mirror is not damaged during operation, because scratched surfaces lower heat-transfer rates.

Instruments operating with liquid helium require elaborate devices (Escaig, 1982) that may not be necessary for preparation of microbial cells.

4.4 Air-dried or freeze-dried, metal-shadowed specimens

Surface reliefs of organisms and cellular appendages, as well as of macromolecules, can best be visualized by application of metal-shadowing techniques. Metal-shadowed specimens contain information about one (the upper) surface only and convey a three-dimensional impression (*Figure 4.13*). Contrast is gained by coating the sample with a thin film of, for example, Pt, Pt–C or W–Ta at angles of between 25° and 50°. When shadowing unidirectionally, metal (or alloy) particles are deposited at the near side, relative to the evaporation source, of a three-dimensional object while the far side ("shadow" region) remains metal free. In consequence, the "shadow" regions appear lighter in the electron microscope than the electron-dense regions where the heavy metal has been deposited. In addition to unidirectional shadowing, rotary shadowing (Heinmets, 1949) can be employed, with the evaporation source held at a constant position while the specimen is rotated. Metal shadowing can be achieved by ordinary resistance evaporation or by electron-beam evaporation (Bachmann *et al.*, 1960; Zingsheim *et al.*, 1970). The equipment required is available from various suppliers (see Appendix B).

It is worth mentioning that the grain size of the deposited metal or alloy usually limits the resolution, which is typically in the region of 2–3 nm. One can generally state that shadowing at lower temperatures (in combination with freeze drying) is superior to evaporation at higher temperatures (Gross *et al.*, 1984), Pt–C to Pt on its own (Bradley, 1958) and W–Ta (which requires an electron beam gun) to Pt–C (Abermann and Bachmann, 1969). A useful comparison of different shadowing techniques is given by Moor (1973). Two examples of typical applications are given by Houwink (1953) and Abram *et al.* (1966). Nowadays, metal shadowing is mainly used in conjunction with critical point drying (used for SEM preparation, see below), freeze drying, fracturing and etching. Although freeze-dried specimens show less artificial distortion or shrinkage than air-dried specimens, the air-drying procedure provides sufficient structural preservation for visualization of rigid structures (e.g. bacterial pili and flagella). The procedure is given in *Table 4.11*.

Carbon-coated grids (see *Table 3.1*) or pieces of freshly cleaved mica are recommended specimen supports. For freeze drying, the sample is

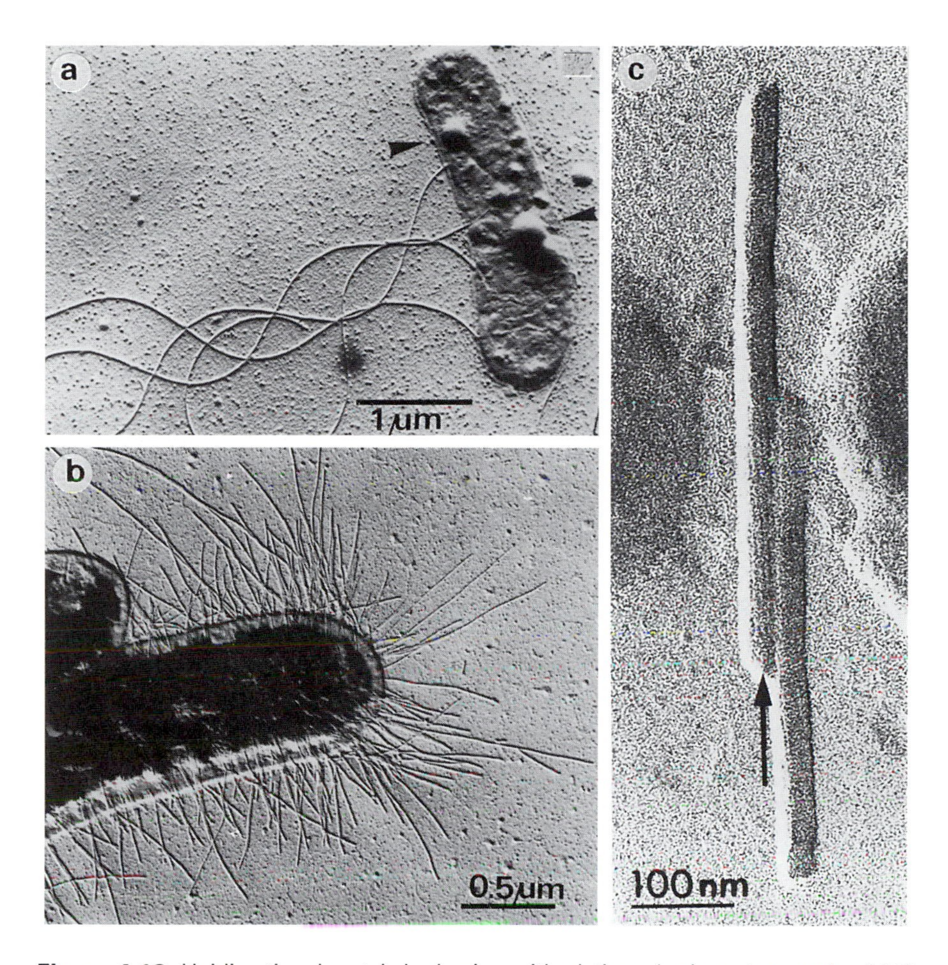

Figure 4.13. Unidirectional metal shadowing with platinum/carbon at an angle of 30°. (a) Flagellated cell of *Ralstonia eutropha,* after unidirectional metal shadowing (original micrograph: I. J. Braks). Note the inclusion bodies (arrowheads) *inside* the cells protruding from the surface. (b) Cell of *Escherichia coli* exhibiting type-I pili (fimbriae) of an *E. coli* cell after unidirectional metal shadowing. (c) Polysheath (defective bacteriophage; Walter-Mauruschat *et al.,* 1977) after unidirectional metal shadowing. Fine structure revealing the arrangement of the subunits (after further image processing) is clearly visible. The arrow marks rows arranged parallel to the length axis.

subjected to plunge freezing in liquid propane, subcooled nitrogen or liquid nitrogen slush (e.g. Robards and Sleytr, 1985) and then transferred to a freeze-drying unit with built-in guns for electron-beam evaporation (*Table 4.12*). Should simple plunge freezing not yield satisfactory results, spray freezing may be a viable alternative. After freeze drying the sample at –80°C (a temperature that is practical with many specimens), the specimen is subjected to shadowing with, for example, W–Ta (Kistler *et al.*, 1977). A comprehensive survey of freeze-drying time/temperature combinations is given by Umrath (1983).

Table 4.11. Air drying and metal shadowing

1. Prepare and check the coating unit including evaporation source(s) and deposit thickness monitor(s). Make sure that you have adjusted the elevation angle to the desired value (20° is a common setting for bacteria). Pump the recipient down and leave under high vacuum
2. Now prepare the specimens: Apply 2–5 μl of sample solution on to a suitably coated grid or piece of mica. Allow the sample to settle for 1 min. Alternatively, float the grid or mica for 20–90 sec on a droplet of sample solution. Gently drain off any excess liquid
3. Wash once or twice with double-distilled water taking care not to wash the sample off
4. Drain off liquid sideways or from below (using filter paper) until dry
5. Transfer the specimens into the recipient of the coating unit and proceed as per user manual to establish a vacuum sufficient for shadowing
6. Shadow specimens unidirectionally (rotary shadowing is preferred with, e.g. nucleic acids because of their otherwise constantly changing orientation relative to the shadowing direction) with W–Ta or Pt–C (coarser grain) to a deposit thickness of about 1–2 nm. Samples on mica need to be additionally coated with a ~10 nm thick layer of carbon at an elevation angle of 90°
7. Vent recipient and remove specimens. Specimens on mica need to be floated off on double-distilled water and mounted on grids. Note that W–Ta-shadowed specimens need to be kept in a vacuum until viewed in the electron microscope to avoid oxidation coupled with a loss of contrast

Table 4.12. Freeze drying and metal shadowing

1. Place a small droplet (~5 μl) of the sample solution on to a suitably coated grid. In cases where the sample solution contains ingredients that could potentially interfere with all subsequent steps (e.g. high concentrations of non-volatile cryoprotectants), one or two washing steps with double-distilled water or filtered isotonic buffer should be introduced
2. After 1 min, blot off the solution (e.g. sideways using two small filter paper triangles) until an approximately 0.1 mm thin liquid film remains. Immediately plunge freeze (or similar) in liquid nitrogen slush (or other suitable cryogen)
3. Transfer the grids on to the precooled (at liquid nitrogen temperature) specimen holder and flange on the precooled (–100°C) specimen table sited inside the recipient. Make sure all specimen manipulations/transfers are carried out in an atmosphere of cool and dry nitrogen gas (i.e. the gas layer settled above liquid nitrogen)
4. Pump down the recipient to a vacuum of greater than 5×10^{-6} Torr and adjust the temperature to –80°C for efficient ice sublimation. Monitor the freeze-drying process by plotting time vs. pressure. Once the curve reaches a plateau (at e.g. 1×10^{-7} Torr), one can be sure that the freeze-drying process is complete
5. The specimens are now ready to be shadowed (see above). The lower the temperature and the better the vacuum, the finer the grain and the higher the resolution. Ideally, keep the shadowed specimen at low temperature, cryotransfer it into the electron microscope and observe at low temperature. In practice, it is sufficient to shadow the specimen at the freeze-drying temperature. Prior to venting (preferably with nitrogen), the temperature is raised to 5–10°C above room temperature while a cold (condensation) trap (at –80°C) is positioned directly above the specimens
6. Transfer the specimens immediately into the electron microscope

4.5 Freeze fracturing and freeze etching

These techniques are used for preparing replicas from frozen surfaces of biological specimens (Robards and Sleytr, 1985) in order to gain information about the three-dimensional structure of the sample as judged from surface reliefs along fracture planes. Fracturing occurs preferentially between the two leaflets of a lipid bilayer and structural inhomogeneities such as inclusion bodies or spores. The applications are numerous and have been reviewed extensively (see, e.g. Holt and Beveridge, 1982). Again, the sample is processed by plunge freezing a sample droplet loaded on to a specimen carrier. The frozen sample is then transferred into a freeze-fracture apparatus as manufactured by Cressington and BAL-TEC (procedure in *Table 4.13*). Very thin-layered samples, for example, monolayers, can be easily fixed by cryo-immersion. With thicker samples, such as pelleted bacterial suspensions, it may be necessary, as a last resort, to apply chemical prefixation (e.g. 2% v/v glutardialdehyde in an appropriate buffer for 3 h) and cryoprotection (e.g. by the gradual addition of glycerol, final concentration 30% v/v, to the sample). Note that glycerol-penetrated specimens cannot be etched and one may therefore wish to consider the use of volatile cryoprotectants (see below). A better way forward is to use as small a volume of the sample solution as possible or to sandwich the sample between two specimen supports (as used for the double replicas; see step 3b in *Table 4.13*).

While at low temperature and in a high vacuum, the specimen is sectioned with a cooled blade (or by the action of a double-replica holder) such that fractured surfaces are exposed. With the double-replica holder, two complementary fracture faces are generated. Cell surfaces *below* the plane of fracturing can be subsequently exposed by controlled sublimation of the surrounding ice. During this sublimation process, which is commonly referred to as etching, about 100 nm of ice is lost per minute (at 2×10^{-6} Torr and –100°C). A replica is formed by evaporating Pt–C or W–Ta, at an angle, on to the frozen surface. This is then stabilized by the deposition of a supporting layer of carbon from above. After deposition of the carbon, the chamber is vented, the replicated specimen is removed, and the remains of the specimen are dissolved from the back of the replica using suitable reagents (e.g. chromic acid or sulphuric acid). After cleaning, the replicas are transferred to grids and can be visualized in the electron microscope without further treatment. W–Ta replicas need to be stored in a desiccator/vacuum chamber since contrast is lost on oxidation. The essential steps of the procedure are presented in *Figure 4.14*.

To avoid confusion in freeze-fracture studies, it is particularly important that the correct replica nomenclature is used as follows (see also *Figure 4.7*). Upon fracturing a membrane by separating the two leaflets, four instead of two surfaces are obtained. The ("old") surface which points towards the extracellular space is called the extracellular surface (ES). It is a surface that has existed prior to fracturing. The same holds true for the "old" surface pointing towards the protoplasm, and this one is called the P surface (PS). The new surfaces, created by fracturing, are called fracture faces, and, depending which leaflet they can be ascribed to, either the E leaflet (also referred to as the E half of the lipid bilayer) or the P leaflet (P half), they are termed E face (EF) or P face (PF). The sequence of (sur)faces approaching from the extracellular space is therefore ES, EF, PF and PS. This nomenclature is applicable to all (sur)faces throughout a cell that can be traced back to either an E or P origin and is therefore not

Table 4.13. Freeze-fracturing procedure (to be used with reference to the instruction manual of the freeze-fracture apparatus)

Day before the experiment

1. Ensure that all equipment to be used is in good working order (vacuum pumps, guns, deposit thickness monitors, quick-freezing apparatus) and that there is a sufficient supply of liquid nitrogen and the other cryogens involved. It is advisable to log the use of different evaporation sources in order to determine readily whether a particular source ought to be changed for the forthcoming experiment. It is advisable to establish guidelines as to how many uses (monitored, e.g. via the total thickness deposited in nm) are allowed before sources have to be exchanged. Note that any monitoring data are source-, angle- and apparatus-specific
2. The sample solution should be as concentrated as possible (barely pipettable)

On the day of the experiment

1. Clean the gold or copper specimen supports (also referred to as specimen discs or specimen mounts) in 50% (v/v) sulphuric acid, followed by double-distilled water (sonication is recommended), and finally acetone, in which they are left until use. Copper supports may, in addition, be pretreated using finishing-grade (400–600) sanding paper. This cleaning procedure is important in so far as it improves the friction between sample and support
2. Turn on cooling water for all equipment that requires cooling (e.g. pumps, guns). Chill the cold trap of the oil diffusion pump, specimen holder, specimen table and knife. Maintain levels of liquid nitrogen. Switch on master switch, rotary pump and oil diffusion pump. Wait ~30 min for the diffusion pump to warm up. Follow the manual to initiate pumping down and cooling down (specimen table, knife). A temperature of around –150°C should be reached

3a. For a single replica: Remove the support from the acetone, allow it to dry, and apply 1 μl of the sample solution so that a pronounced convex meniscus protrudes from the surface of the support. Quick freeze, mount into the precooled specimen holder and transfer into the recipient. Pump down

OR

3b. For double replicas: Remove two supports suitable for double replicas from the acetone and allow them to dry. Apply 1 μl of the sample solution on to the first carrier and place the second one upside-down on top of the first one, pressing firmly together. Any excess of liquid should be blotted off using filter paper. Quick freeze this sandwich assembly and mount into a precooled double-replica holder. Transfer into the recipient and pump down

4. Fracturing, etching and shadowing: Raise the temperature to a value suitable for freeze-etching (at a pressure of $\leq 10^{-7}$ Torr, a suitable temperature is e.g. –98°C) and ensure that the knife is cooled down to < –190°C. Once a temperature of –98°C is reached, wait a further 15 min in order to avoid temperature gradients between the specimen table and specimen supports. Once the temperature is constant, fracture the specimen with the cold knife or by flipping open the double-replica holder

   Immediately position the knife as a cold trap directly above the specimen. The specimen is now etched, i.e. ice sublimes. The time is monitored exactly so that the desired effect is achieved, e.g. deep-etching (> 30 min), regular freeze etching (typically 3–5 min) or predominantly freeze-fracturing (≤ 30 sec). In the latter case, the etching time is kept to a minimum

   N.B. The times provided are only guidelines for the above temperature–pressure combination

   As soon as the estimated time has elapsed, the cold trap is removed, and the specimens are immediately shadowed, first with a contrast-imparting metal (e.g. 1 nm thick Pt–C at 45°C) and then with carbon (10–20 nm thick at 90°C). Thus, it is good practice to ensure, before beginning the process, that the equipment for shadowing is ready for immediate use. Fracturing, etching and shadowing ought to be seen as a group of activities carried out in quick succession, unless deep-etching is desired

(Continued)

5. The specimens are now ready to be removed from the recipient (follow the instruction manual of your particular freeze-fracture apparatus) and cleaned. This final step prior to electron microscopy can be extremely unforgiving if carried out carelessly. Published protocols vary. The following has worked well with bacteria: float the replicas off on 70% sulphuric acid and leave for 12 h. If there are problems with this first step, float off on double-distilled water and then on solutions containing increasing concentrations of sulphuric acid. After a few hours one should have arrived at the final solution, on which the replica is left for 12 h. Using a platinum loop, transfer the replica on to either a 6% (w/v) sodium hypochlorite solution or 20% chromic acid and leave for 1–2 h. Next, clean the replicas in double-distilled water (3 x 20 min). Transfer on to a grid and, using pointed filter paper, carefully blot off any liquid from the replica. Grids can be used coated or uncoated, but one should ensure that the surface is hydrophilic (e.g. by glow discharging, see Section 3.2.1). With coated grids, there is generally less chance of making the replicas burst during the blotting step

Having finally arrived at the stage where one is observing the replica in the electron microscope, and having done everything to avoid artefacts, here is a quick check-list how to recognize an artefact-free freeze-fracture replica:

Look out for

- a very fine grain of the shadowing material
- an overall crisp contrast of the replica
- a good correlation between the complementary fracture faces (P and E faces, where applicable)
- smooth contours around cells and organelles
- a fine granular texture of the frozen medium around the sample (non-etched preparations only)
- an even or mosaic distribution of particles when dealing with fracture faces of membranes

---

disturbed by any membrane fusion events. With bacteria, however, one can encounter problems, because the cells are surrounded by, for example, cell walls, outer membranes and surface layers, that is, by entities which themselves can undergo fracturing. Therefore, in cases where the use of the P/E nomenclature by itself is not sufficient, one should attempt to supply additional descriptions that appropriately reflect the individual circumstances. Varga and Staehelin (1983) have presented data on the membrane system of *Rhodopseudomonas palustris* where the P/E nomenclature is used in conjunction with other descriptive labels.

In the following, typical causes of artefacts are described, together with strategies for how to avoid them. This section should therefore always be considered essential reading when planning a freeze-fracture experiment:

- If cells are frozen at too low speeds ($<100°C\ sec^{-1}$), intracellular water may crystallize and destroy any fine structure (*Figure 4.15a*). However, the overall shape of the cell may not be influenced. The use of volatile (no interference with the etching process) cryoprotectants is the best remedy in the absence of means of improving the freezing rate. Examples of volatile cryoprotectants are methanol and ethanol. Note that inserting a sample too slowly into a good cryogen may lead to as disappointing a result as rapid injection into a poor cryogen.
- When fracturing with a knife, the knife edge should not touch the fracture surface so that knife marks (wheel cart tracks) on the surface are avoided (*Figure 4.15b*). The risk of knife marks may be reduced by avoiding sweeping action, using sharp knives with

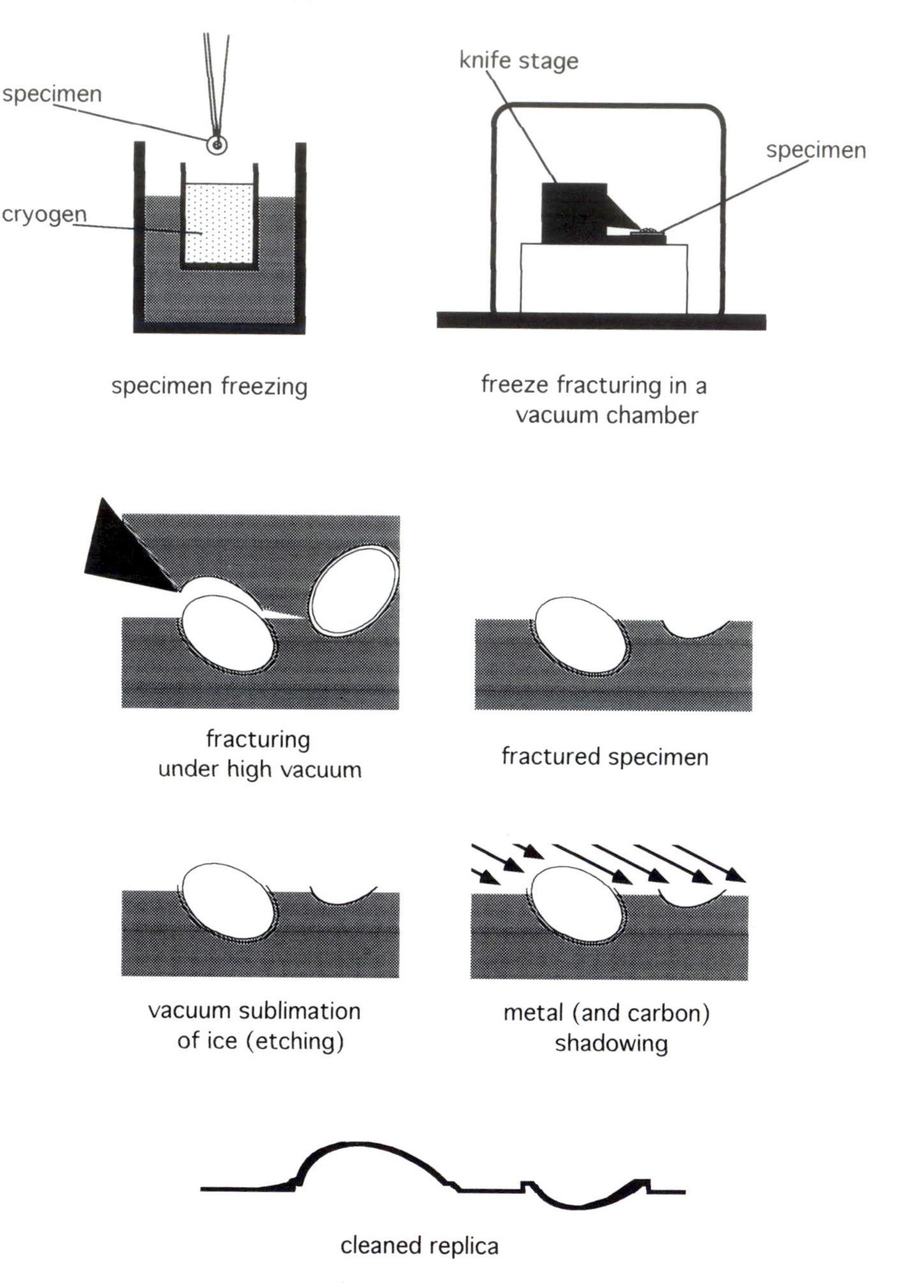

Figure 4.14. Basic steps of the freeze-fracturing procedure (according to Shotton and Severs, 1995). See Section 4.5 and *Table 4.13*.

non-jagged edges and by increased cutting speeds. A small knife advance can be used to avoid steps on the fracture surface.

- Plastic deformation can occur with materials that are still deformable at low temperature. Examples of such materials are polyphosphate and poly-β-hydroxy butyric acid (PHB) inclusions.

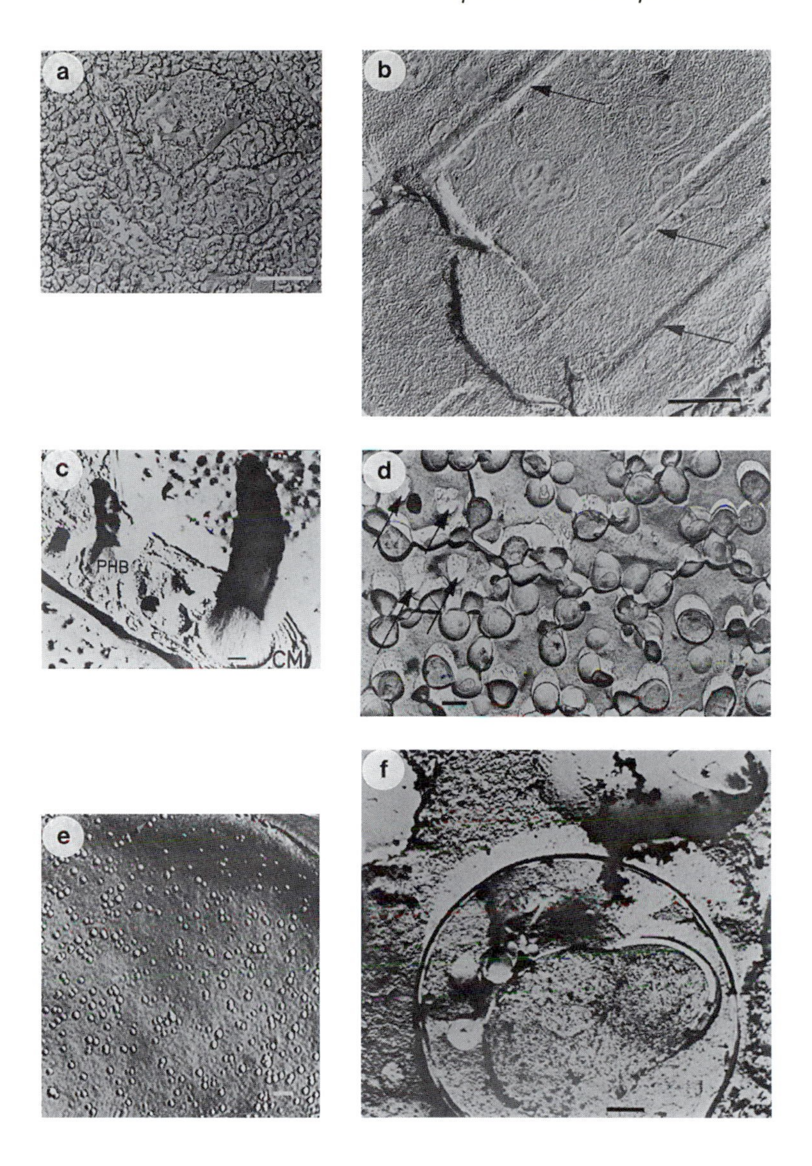

Figure 4.15. Typical freeze-fracture/etching artefacts. In (a), cells have been frozen at too low a speed, leading to the crystallization of intracellular water and a complete destruction of any fine structure. (b) Upon contact between the fracture surface and the knife edge, exposed specimen structures (here shown with yeast cells) become indistinct and knife marks may be observed caused by a jagged knife edge (marks are highlighted by arrows). (c) Longitudinally fractured *Nitrobacter winogradskyi* showing deformed PHB inclusion bodies (CM = intracytoplasmic membranes). In (d), *Staphylococcus areus* cells are observed that have been etched too deep causing some of the cells to drop off the replica (arrows). Panel (e) shows the effect of condensation due to water vapour. In this case, the temperature was too low for the pressure used. (f) This electron micrograph demonstrates a likely outcome when contaminated distilled water is used for cleaning the replicas. Scale bars correspond to 1 μm in (a), (b) and (d), 100 nm in (c) and (e) and 200 nm in (f). This figure has been assembled from previously published material with the kind permission of S. Böhler BAL-TEC AG, Liechtenstein, accrediting H. Moor, A.P van Gool, K.G. Lickfeld and P. Roehrlich, and the BALTEC EM-Laboratory for original micrographs.

PHB inclusions can be pulled out of the fracture surface into conical protrusions with a typical appearance (*Figure 4.15c*). This type of artefact is quite common (Sleytr and Robards, 1977) and must, therefore, always be considered when interpreting images of microorganisms containing polymeric inclusions.

- Etching too deeply can cause the ice level to decrease to such an extent that cells become completely exposed and lie loosely on the surface. These cells may be removed during subsequent preparation steps, leaving holes in the replica (*Figure 14.15d*). This artefact can only be avoided by precisely monitoring the temperature and etching time. Etching (sublimation) rates are provided in *Figure 4.16a*.
- Any residual water vapour that condenses on the specimen prior to replication gives rise to the formation of ice crystals which then look like intramembane particles (small blobs; *Figure 4.15e*). Water vapour may arise (i) generally during the sublimation of ice (etching, see above), (ii) from working at temperatures/pressures that favour condensation (see *Figure 4.16b*), and (iii) from the shadowing sources. Condensation artefacts generally — not only caused by water vapour but also by other condensing gases — can be avoided

  — by carefully selecting the temperature (at a given pressure);
  — by protecting the specimen sufficiently via a cold trap, like the cooled knife and cold shroud (the temperature of which must be below that of the specimen!) OR, in cases were sufficient protection cannot be established, by immediately coating the freshly fractured surface (any etching effect will be negligible in this case);
  — by ensuring that a good, clean vacuum exists around the specimen (diffusion pumps should be fitted with cold traps, and these ought to be used; rotary pumps should be run on gas ballast at regular intervals and for sufficient periods of time), and
  — by thoroughly degassing the electron beam or resistance-heated evaporation sources (guns) prior to each experiment.

- Inadequate (little contrast and no detail) or excessive shadowing (fine details are masked) can be avoided by utilizing a deposit-thickness monitor (for standard values, see step-by-step protocol in *Table 4.13*). If commercial film thickness measuring devices are not available, one can improvise e.g. with pieces of filter paper or gold foil. The former will increasingly darken (make sure part of the paper is folded up to create an area inaccessible to shadowing for reference) and the latter will assume thickness-dependent interference colors. Once optimum conditions have been found, *all* parameters should be documented with the utmost precision.

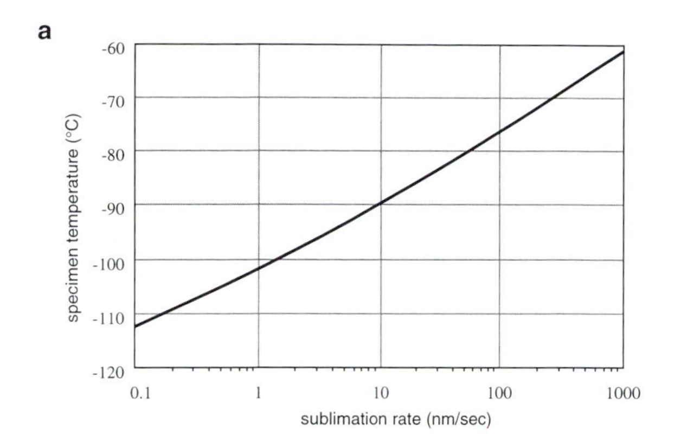

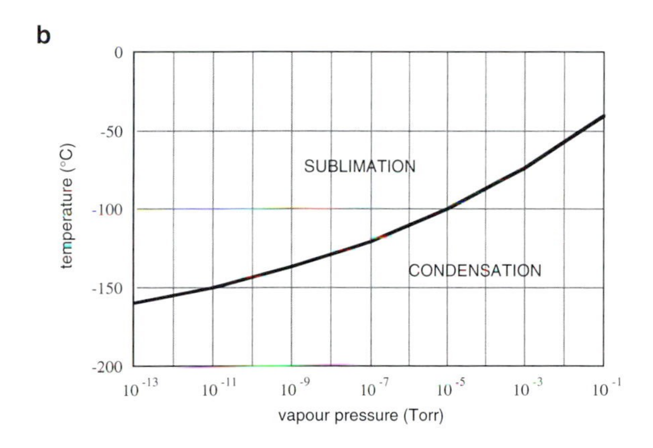

Figure 4.16. (a) Diagram showing the relationship between sublimation/etching rate and temperature, assuming an appropriate pressure. (b) Diagram of the saturation pressure of water.

- Superficial cleaning of the replicas or the adoption of unsuitable cleaning procedures and chemicals (e.g. insufficient purity) will invariably lead to smudges across an otherwise perfectly good specimen. A specimen cleaning procedure is provided in the step-by-step protocol. If contaminated water is used during the final cleaning steps, the result may be extremely disappointing with dark patches covering large areas of the replica (*Figure 4.15f*). As a rule: always use double-distilled water.

4.6 Sample preparation for thin sectioning

Ultra-thin sections of objects provide direct insight into their internal structure without the requirement to assess images along fracture

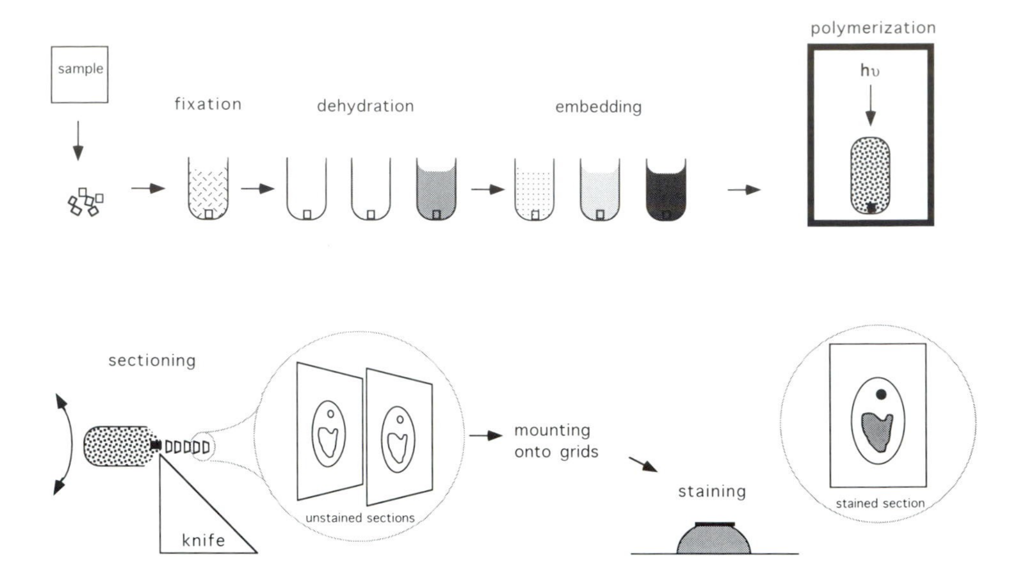

Figure 4.17. Basic steps for resin embedding and ultra-thin sectioning of specimens.

planes and surfaces as revealed by the techniques described above. Normally, cells, subcellular particles or vesicles are visualized in thin-section studies. Numerous different procedures for the fixation, dehydration, embedding and staining of biological specimens have been developed in order to achieve preparations with minimum artefacts (see, e.g. Hayat, 1981). Though various preparation procedures may be applied, the important basic steps are identical for all procedures and are summarized in *Figure 4.17*.

4.6.1 Chemical fixation

Chemical fixation of proteins and lipids is necessary when dehydration in organic solvent and resin embedding are applied. Widely used fixatives are described in *Table 4.14*.

Cells from numerous bacterial species (such as *Escherichia coli*, *Pseudomonas* and *Bacillus* strains) may be chemically fixed in unbuffered aqueous solutions without any obvious loss of fine structure compared to cells fixed in buffered solutions. Nevertheless, for most specimens, more careful handling is recommended.

Fixatives such as osmium tetroxide cause a drop in pH in the sample. Therefore, appropriate buffering of the fixation steps is recommended in order to avoid artificial change in the specimen caused by acidification.

Sensitive organisms such as spirochaetes or marine microorganisms should be fixed in a physiological medium to prevent swelling of the cells or other damage. The optimal osmolarity for these specimens has to be determined by trial and error. It has to be kept in mind that the osmotic

Table 4.14. Fixatives for biological specimens

Fixative	Main effect	Solution (buffered, if not indicated otherwise) and special features
Glutardialdehyde (Sabatini *et al.*, 1963)	Cross-linking of proteins	0.5% (v/v)–5% (v/v) 3% (v/v) for most procedures inclusion of formaldehyde (up to 4% w/v, freshly prepared from paraformaldehyde) is useful, especially when rapid fixation of specimens is necessary since it improves the penetration. A 0.2% (v/v) glutardialdehyde/0.3% (w/v) formaldehyde solution is suitable for fixation of bacterial cells in immunocytochemisty
Acrolein (Luft, 1959)	Cross-linking of proteins	To be used as alternative to the other aldehyde fixatives only when rapid fixation is of paramount importance
Osmium tetroxide (Millonig, 1961)	Polymerization of hydrocarbon chains (lipids)	1% (w/v) solution (inclusion of 0.5–1%, w/v, potassium ferricyanide may provide enhanced staining of membranes)
Permanganate (Luft, 1956)	Fixation of biological membranes	1% (w/v) aqueous solution only membranes are preserved, other cell components are damaged or removed

pressure of the solution changes when fixatives are added; 0.1 M phosphate buffer has an approximate osmolarity of 210 mOsm, and 4% (v/v) glutardialdehyde raises osmolarity to 710 mOsm. For fibroblasts, for instance, an appropriate fixation medium consists of 90 mM sucrose, 10mM $MgCl_2$, 10 mM $CaCl_2$, 3% (v/v) glutardialdehyde and 5% (w/v) paraformaldehyde in a buffered medium (0.1 M sodium cacodylate buffer, pH 7.0). For many organisms, the original growth medium is the ideal fixation medium. After fixation of the specimen in osmium tetroxide (OsO_4), the organisms are no longer osmotically active (even after fixation with aldehydes, obvious osmotic damage of the sample is rare). Therefore, osmotic stabilization is no longer necessary after this step. Standard fixation of bacterial cells is carried out in an appropriate buffer (or buffered culture medium) including up to 5% (v/v) glutardialdehyde for 1–2 h at room temperature. After washing, the cells are resuspended in a small aliquot of agar. When cut into small blocks of about 1 mm^3 the sample is easier to handle than cell suspensions. Moreover, the agar protects the cells against mechanical disruption.

4.6.2 Dehydration, infiltration and embedding

Removal of free water from the specimen before embedding in a suitable resin (see *Table 4.15*) is achieved by incubation of the specimen in a graded series of organic solvents (methanol, ethanol or acetone). The dehydration time should be as short as possible in order to reduce

Table 4.15. Resins for standard embedding procedures

Resin	Constituents[a]	Polymerization conditions
Epon (Kushida, 1967)	Epon 812 (16 g) Dodecenyl succinic anhydride (8 g) Methyl nadic anhydride (8.7 g) Benzyl dimethylamine (0.4 g)	24–72 h at 60°C
Araldite (Glauert *et al.*, 1956)	Araldite CY212 (29 g) Dodecenyl succinic anhydride (24 g) 2, 4, 6-Tridimethylaminomethyl phenol (0.5 g)	5 h at 45°C and 12 h at 60°C
Spurr (1969)	Vinylcyclohexene dioxide (ERL 4206) (10 g) Diglycidyl ether of polypropylene glycol (6 g) Nonenyl succinic anhydride (26 g) Dimethyl aminoethanol (0.4 g) (a mixture of lowest viscosity is composed of hexenylsuccinic anhydride and Araldite RD 2 instead of nonenyl succinic anhydride and the diglycidyl ether)	8–12 h at 70°C
Nanoplast (Frösch and Westphal, 1989)	Hexamethylolmelamine methyl ether (MME 7002, 70% v/v in water) (10 g) *p*-Toluenesulphonic acid (acid catalyst B 52) (0.2 g)	Desiccated for 48 h at 40°C 48 h at 60°C

[a]Quantities of the components may be varied in order to obtain harder or softer polymerized blocks.

shrinkage and keep extraction of tissue components at a minimum, but should allow complete removal of water from the sample. For rigid specimens (filamentous fungi, biofilms or tissue), longer dehydration times are recommended. A typical dehydration procedure for unicellular organisms (e.g. bacteria and microalgae) involves incubation of fixed samples in a graded acetone series for 3–15 h (see *Table 4.16* for details). After dehydration, the sample is infiltrated by an embedding resin (see *Table 4.15*). A vast number of variations from the protocol as given below have been described (though systematic investigations concerning the effect of variations in the embedding procedure are rare). For most unicellular bacteria, wide variations in fixation time, concentration and temperature during fixation, dehydration time (as long as it is above a certain minimum) and temperature have little or no effect on preservation of ultrastructure (although enzyme activity and antigenicity may be affected, see respective chapters). Conditions should, of course, be kept constant for samples that will be subject to direct comparison of, for example, cell volume or dimensions of subcellular compartments, but may be adapted to suit other requirements (e.g. sampling time, storage or shipping). The time required for infiltration of bacteria with resin is dependent on resin viscosity (and, therefore, indirectly, temperature) but is, in general, 3–18 h.

The resins Epon, Araldite and Spurr are soluble in acetone, but not in ethanol. If ethanol is employed as a dehydrating agent, a transitional solvent, propylene oxide, must be used. Infiltration of the embedding medium occurs by gradual replacement of the solvent by the liquid resin. A typical embedding protocol for unicellular organisms in Spurr resin

Table 4.16. Outline procedure for chemical fixation and embedding of samples in Spurr resin

1. Wash cells by filtration (recommended for filamentous cells like moulds or certain cyanobacteria) or centrifugation (15 min at 2000 *g* for most unicellular organisms). Appropriate buffers for the washing procedure are 100 mM phosphate-buffered saline (PBS), pH 7.0–7.5, cacodylate buffer, pH 7.0–7.5, or 100 mM HEPES pH 7.0–7.5 containing 2 mM $MgCl_2$; 100 ml of bacterial culture (OD_{540} ~1) provides enough cells for one preparation.
   For some fragile organisms, that would not withstand the washing procedure, the chemical prefixation (see step 2) should be performed in the culture medium prior to the washing procedure
2. (Prefixation). Prefixation is very useful when the sample has to be stored or transported. The fine structure of prefixed cells is preserved for several months. If prefixation is not necessary, proceed with step 4.
   Add glutardialdehyde (usually available as a 25% v/v solution) to a washed cell suspension or directly to the liquid culture to a final concentration of 0.5% (v/v) and incubate for 90 min at room temperature. Remove the glutardialdehyde solution by filtration or centrifugation of the cell suspension
3. The resuspended pellet (in buffer, see step 1) must be refrigerated. Do not freeze!
4. The second chemical fixation step is performed in 3% glutardialdehyde. If transportation or storage of the sample is not necessary, the prefixation may be omitted. Add glutardialdehyde solution to the sample to a final concentration of 3% (w/v) and incubate for 2–3 h at room temperature. Wash the pellet in buffer three times (see step 1). Resuspend the pellet in a *small* volume of buffer (just enough to permit mixing with molten agar according to step 5)
5. Melt agar in buffer at 1.5–2% and keep at 45°C. Add 1–1.5 volumes of agar to one volume of pellet and mix rapidly. Incubate the agar suspension at room temperature or in an ice bath until the agar is solid. Remove the agar block from the centrifuge tube and cut it into small, approximately 1 mm^3, pieces. Transfer the cubes to small glass vials. The small cubes are easy to handle during the subsequent fixation and embedding procedures
6. Wash the cubes once with an appropriate buffer to remove small pieces of agar and free floating cells. Since the cubes are too large to slip into a pipette tip, no centrifugation or filtration is necessary for this procedure or for the subsequent incubation steps
7. Incubate in 1% (w/v) osmium tetroxide solution for 1 h at room temperature (the incubation time may be extended to up to 4 h to enhance staining of the sample). Wash the cubes (now a dark brown colour) at least three times with distilled water
8. Dehydrate the sample by incubating the cubes in 10% (v/v), 30%, 50%, 70%, 90% and 100% acetone solutions (10–15 min per step, 0°C [ice bath] incubation temperature, acetone solution prepared in distilled water). Improved staining of the specimens is achieved, when the 70% acetone incubation is replaced with one in 2% (w/v) uranyl acetate in 70% acetone (3–12 h). Addition of $CuSO_4$ or $CaCO_3$ as desiccant to the 100% solution some hours prior to use ensures it is water free. Avoid contamination of the sample with desiccant. Repeat the 100% acetone incubation step twice
9. The subsequent steps are performed at room temperature. Add 1 volume of Spurr resin to 2 volumes of 100% acetone containing the samples. Mix and incubate for 30 min. The infiltrated agar cubes sink to the bottom of the vial. Add 2 volumes of resin to the sample and incubate for 1.5 h. Replace the resin–acetone mixture with pure resin and incubate for 8–16 h. Replace the resin with fresh resin and incubate for 1–3 h
10. Transfer single cubes into small gelatin or Beem™ capsules that are full of fresh resin. The cubes should sink to the bottom of the capsules after some minutes. De-gas the capsules in a vacuum chamber (the vacuum provided by a one-stage rotary pump is sufficient) for 5–10 min until the development of gas bubbles stops
11. Polymerize the capsules at 70°C for 8 h

Table 4.17. Outline embedding procedure for Nanoplast resin

1. Prepare the Nanoplast resin according to the supplier's instructions shortly before use (10 g MME 7002 and 0.2 g B 52 are recommended for samples that are appropriate for the preparation of conventional ultra-thin sections)
2. Prepare the sample according to steps 1–4/5 in *Table 4.16*
3. Place a small sample (after step 4) or a small agar cube (after step 5) in a Beem™ capsule, filled to a depth of approximately 5 mm (not more, because otherwise the desiccation procedure has to be extended) with Nanoplast resin
4. Place the capsule in a desiccator containing silica gel and incubate for 2 days at 40°C. Continue incubating the specimen at 60°C (without further desiccation). If the resulting resin block is too soft, incubate the capsule at 80°C for 5–10 h. During the embedding and polymerization process, the resin volume shrinks markedly
5. If the resulting resin block is difficult to handle because of extensive shrinkage, embed the capsule in a conventional resin. After polymerization, the surrounding resin provides a mechanical support for the Nanoplast resin

(including an appropriate fixation and dehydration procedure) is presented in *Table 4.16*.

Nanoplast resin embedding procedure. Embedding in the highly water soluble Nanoplast™ melamine resin does not require dehydration of the sample by organic solvents (Frösch and Westphal, 1989; *Table 4.17*). This may be advantageous for cells that suffer ultrastructural change as a result of dehydration with organic solvents. Though conventional fixation with glutardialdehyde may be performed, treatment with osmium must be omitted. Pretreatment of the sample before resin embedding by this method is short and simple, but the desiccation and polymerization procedures are time consuming.

Embedding media for immunocytochemistry. The embedding media described above are not recommended for immunocytochemical localization of cellular constituents. Suitable resins for immunocytochemical studies are given in *Table 4.18*.

Low temperatures (–35°C to –70°C) during the dehydration and infiltration steps result in excellent preservation of structure, antigenicity and enzyme activity. Therefore, Lowicryl resins are the most widely used embedding media when these features are desired. A detailed protocol for an embedding procedure using Lowicryl K4M resin is given in *Table 4.19*.

Some samples (especially dilute suspensions) are difficult to detect in polymerized resin, because the intense osmium stain is absent. Embedding a visible marker (at best included in the agar medium; see step 5 of *Table 4.16*) to allow an easy localization may be helpful, but the marker should not interfere with the sample and must be insoluble in the resin. Fine carbon particles from indian ink co-entrapped in the agar medium are helpful (centrifuge the ink before use and resuspend in the buffer that is used during the embedding procedure). The stain should result in no more than a light grey appearance of the sample. Higher concentrations would prevent sample polymerization by UV light! It has

Table 4.18. Resins suitable for immunocytochemistry

Resin	Special conditions for embedding and polymerization	Special properties
Glycol methacrylate (Leduc and Bernhard, 1967)	Water (instead of organic media) as solvent; polymerization in UV light possible	Sections relatively unstable to the electron beam
London resins (LR)	Dehydration in methanol or acetone/ ethanol; complete dehydration of the sample not necessary; polymerization in UV light possible	Very low viscosity (infiltration of rigid specimens)
LR White (Newman and Hobot, 1987)	Use at room temperature	
LR Gold (Shires *et al.*, 1990)	Use at low temperature (–4 to –25°C)	
Lowicryl resins (K4M, HM 20; Carlemalm *et al.*, 1982)	Dehydration in methanol or ethanol, deep-temperature embedding, polymerization in UV light possible	Low viscosity at low temperatures, excellent preservation of structure and antigenicity of the sample
K4M	Embedding temperature as low as –35°C, polar resin (residual amounts of water in the sample are tolerable)	
HM 20	Embedding temperature as low as –70°C, non-polar (no residual water is tolerated)	

to be borne in mind that the carbon particles produce a background of irregular dark spots in the ultra-thin section.

The use of Lowicryl resin requires cooling devices capable of producing an ambient temperature of –35°C or lower for a long period of time for optimal results. At this temperature, denaturation of proteins in organic solvents is considerably reduced. However, the embedding and polymerization procedures may be performed at 0°C or at room temperature. During polymerization at room temperature, small gas bubbles can be generated which may give rise to problems during ultra-thin sectioning. Increased gas production occurs when dibenzoyl peroxide is used in conjunction with heat polymerization (as proposed by the supplier as an alternative polymerization method). This method is not recommended.

Freeze substitution. Cryofixation of the samples can be coupled with (low-temperature) resin embedding procedures via freeze substitution. This technique is used when the conventional dehydration procedure is to be avoided, for example, when water-soluble substances (small ions) should not be removed, but the frozen water has to be replaced with an anhydrous solvent, typically followed by resin infiltration. With the aid of freeze substitution, the advantages of sample vitrification are combined

Table 4.19. Outline embedding procedure for samples in Lowicryl K4M

1. Prepare a fresh 5–10% (w/v) formaldehyde solution from solid paraformaldehyde by suspending paraformaldehyde in distilled water and heating to 80°C. Slowly add drops of 1 M sodium hydroxide until the solution becomes clear
2. Wash cells by filtration (recommended for filamentous cells like moulds or certain cyanobacteria) or centrifugation (15 min at 2000 *g* for most unicellular organisms). Use 100 mM PBS (pH 7.0–7.5) for the washing steps. Cacodylate buffers are not recommended, because they may reduce the antigenicity of the specimen. 100 ml of bacterial culture (OD_{540} ~1) provides enough cells for one preparation.
   For some fragile organisms, that would not withstand the washing procedure, the chemical fixation (see step 3) should be performed in the culture medium prior to the washing procedure
3. Add glutardialdehyde and formaldehyde solutions to a washed cell suspension or directly to the liquid culture (step 2 is omitted) to final concentrations of 0.2% (v/v) and 0.3% (v/v), respectively, and incubate for 90 min in an ice bath with gentle shaking. Centrifuge 3 times for 15 min at 2000 *g* in PBS containing 10 mM glycine
4. Embedding in molten agar as described in step 5 of *Table 4.16*
5. Dehydrate the sample by incubating the agar cubes in 15% and 30% methanol (15 min per step) at 0°C, in 50% methanol (15 min) at –20°C, in 70%, 90% and 100% methanol (30 min per step) at –35°C
6. Add 1 volume of pre-chilled Lowicryl resin per volume of methanol and incubate for 1 h at –35°C. Add 0.5 volume of resin per volume of mixture and incubate for 1 h at –35°C. Replace the resin–methanol mixture with fresh resin and incubate overnight at –35°C. The infiltrated agar cubes sink to the bottom of the vial. Replace the resin with fresh resin and incubate for 2 h at –35°C
7. Transfer the agar pieces into gelatin capsules. The cubes will sink to the bottom of the capsule within 30 min. Polymerize for 40 h at –35°C and 3 days at room temperature (may prove unnecessary) with UV light (370 nm, 15 W, distance: 30–40 cm. To provide diffuse illumination a right-angle reflector is suspended below the UV lamps)

with procedures that result in resin-embedded sections that are easier to handle and store than frozen specimens. Ultra-thin sectioned specimens prepared by freeze substitution are depicted in *Figure 4.18*. Besides resin embedding, samples may be processed by subsequent freeze drying or critical-point drying methods. See Hippe-Sanwald (1993) for review, and Paul and Beveridge (1993) for application examples.

Samples are plunge frozen and then transferred to a freeze-substitution system in which they are gradually warmed to a temperature that permits resin embedding (e.g. –40°C for subsequent deep temperature resin embedding or ambient temperature for conventional resins). During the freeze-substitution process, the ice in the sample is replaced by an organic solvent. For subsequent X-ray microanalysis of the specimen, water has to be removed completely by the use of water-free absolute methanol (dried over a molecular sieve and with the addition of a molecular sieve during freeze substitution). Several authors recommend chemical fixation (glutardialdehyde, osmium tetroxide) during methanol infiltration in order to stabilize the cellular constituents after the ice has been removed (see Robards and Sleytr, 1985, for survey). However, this process may destroy the superior structural preservation that has been achieved by rapid freezing (Sjöstrand, 1990).

A simple set-up for cryofixation and freeze-substitution devices is described by Sitte (*Figure 4.19*; see Robards and Sleytr, 1985; Roos and

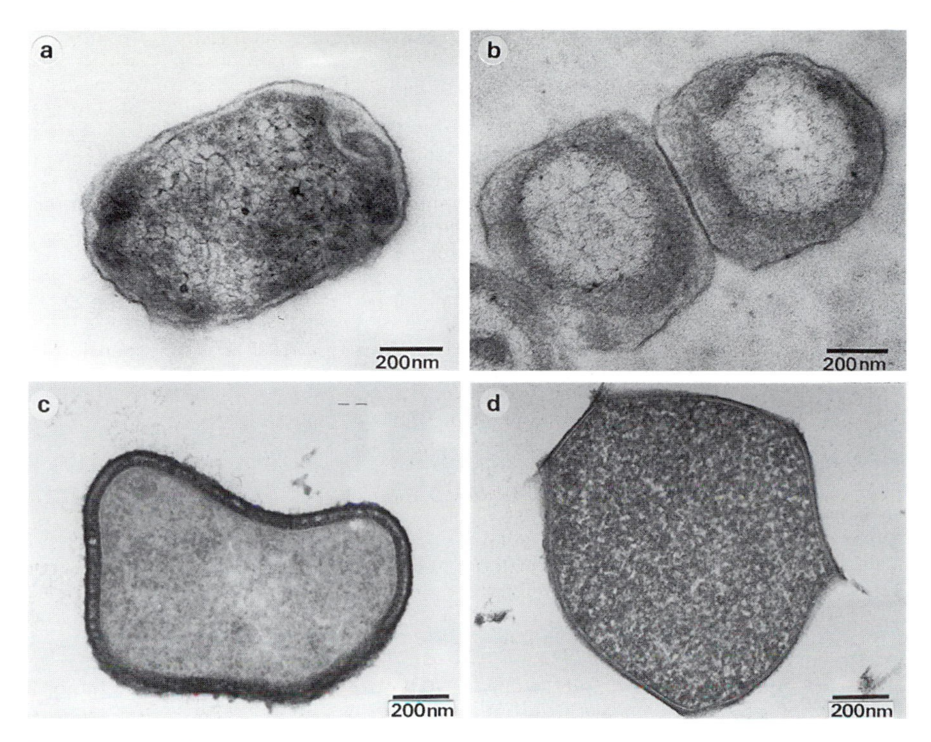

Figure 4.18. Ultra-thin sections from specimens prepared by conventional and deep temperature embedding methods. (a) *Thiosphaera pantotropha* cells after chemical fixation and embedding in Spurr resin. (b) The same specimen as in (a) after chemical fixation and processing by deep temperature embedding in Lowicryl K4M resin. Note the difference in appearance of the central nucleoid. Original micrographs in (a) and (b): Y. Wüstefeld; specimens prepared by S. Ullmann. (c) *Pyrolobus fumarii* and (d) *Thermococcus chitonophagus*, both prepared by rapid freezing and subsequent cryosubstitution revealing an excellent preservation of cellular shape (no artificial wrinkles or shrinkage). Original micrographs: R. Rachel. See also Huber *et al.* (1995) and Blöchl *et al.* (1997).

Morgan, 1990). A commercially available device is based upon the instruments developed by Sitte and Edelmann (Sitte, 1984; CS-Auto, Reichert-Cambridge Instruments).

A simple freeze-substitution device may be constructed using normal workshop facilities: An aluminium container is placed over liquid nitrogen (not submerged) in a dewar. Cooling is provided by the nitrogen gas surrounding the container. The minimum temperature of this device is approximately –80°C. Higher temperatures are reached by adjusting a thermosensor/heater circuit. The bottom of the metal container is insulated to prevent unnecessary nitrogen loss. The container is filled with a substitution medium that is continuously agitated by a stirrer. Frozen specimens are placed on a metal grid inside the container. More elaborate devices are available based on the same construction principle. A standard protocol for freeze substitution of bacterial cells is detailed in *Table 4.20*. Prior to this procedure cells are chemically fixed with 0.2% paraformaldehyde and 0.3% glutardialdehyde (may be omitted, if chemical fixation is

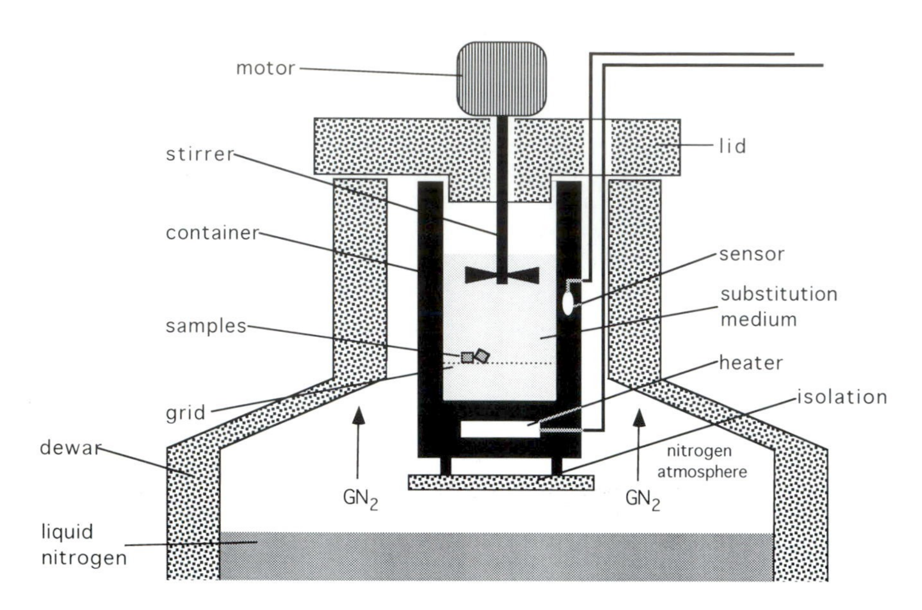

Figure 4.19. Freeze-substitution apparatus (see Section 4.6.2).

Table 4.20. Freeze substitution of samples

1. Fill the container of the freeze-substitution device with absolute methanol and adjust the temperature to –80°C. Allow the system to equilibrate (continuous loss of nitrogen in simple devices may cause an undesired temperature gradient)
2. Transfer the sample from liquid nitrogen (see *Table 4.9*; plunge freezing) to the pre-chilled methanol
3. Allow substitution to occur for 18 h at –80°C followed by gradual warming to –40°C over 60 h
4. Transfer the sample to a refrigerator held at –40°C in the chilled metal container and then proceed from step 6 in *Table 4.19* (embedding in Lowicryl resin). Though the sample itself may be completely invisible in the embedding resin, the support (grid or filter paper) allows the sample to be localized

not desired) and subjected to plunge freezing as described in *Table 4.9*. The cells are then embedded in Lowicryl resin (see above).

4.6.3 Ultramicrotomy

Trimming. Sections of specimens embedded in resin, as described above, should be between 60 and 90 nm thick to allow proper imaging in the electron microscope. Therefore, the samples are subjected to trimming and ultra-thin sectioning. Trimming of the (resin embedded) samples is done in order to obtain small, flat-topped pyramids with a face of about 0.2 mm^2. Trimming of the firmly held resin block may be performed with the aid of a sharp razor blade or, more accurately, with a rotating milling cutter (e.g. a Reichert specimen trimmer with a diamond cutter). The flat top of the pyramid (mesa) should be as smooth as possible. Its shape is of the utmost importance: ribbons of sections (e.g.

for serial sections) are produced most easily when the trimmed surface is a trapezium. A square is also usable, but unsuitable for production of ribbons more than four or five sections long.

Knives. Sectioning of the sample is done in an ultramicrotome with the aid of glass or diamond knives. The former knives are fully sufficient for most standard sectioning procedures (including cryo-ultramicrotomy); the latter are particularly useful when sectioning hard objects, for numerous serial sections or for very thin sections. Glass knives are produced by cleaving square glass pieces (broken from a glass strip) into two triangular knives. Making glass knives with glazier's pliers requires some experience; therefore, special knife makers are used to produce glass knives for standard and cryo-ultramicrotomy with high reproducibility. Knife quality is essential for proper sectioning and needs to be carefully checked. Important features of knife quality are summarized in *Figure 4.20*. Normally, 40 cm long glass strips (6 mm thick) are used to produce 16 squares with right-angled corners. By breaking the squares slightly off the diagonal axis, two usable knives are obtained: the square is scored in such a way that the scores stop at some distance from the corner. When pressure is applied under the score, the fracture extends towards the corners of the square, following the line of the score. At each end of the score, the fracture deviates from the line of the score to curve away from the corner towards the edge of the square ("free" break). This results in the real included angle of the knife being somewhat greater than the angle of the score. The real angle of the knife increases as the score is moved further from the diagonal, that is, when the counterpiece (of the *opposite* knife!) is larger. A counterpiece with a width of 0.4 mm with an adjusted knife angle of 45° results in a real angle of 55° (±2°) at the cutting edge. Increasing the width of the counterpiece by adjusting the knife maker accordingly results in larger angles, while decreasing it reduces the angle to values approaching 45°. The smallest knife angles lead to very sharp, but brittle cutting edges. For sectioning of ultra-thin sections from resin-embedded samples, knives with counterpieces of 0.3 mm–1 mm give the best results. Knives with very small counterpieces (0.1 mm and smaller) are recommended for cryosectioning. The overall appearance of the knife edge and the length of the useful cutting edge vary as indicated in *Figure 4.20d*. By inspection of the cutting edge in dark-field illumination under a stereomicroscope, flaws or frills are easy to detect.

For sectioning resin-embedded samples, a small fluid-filled (normally with particle-free water) trough is attached around the knife edge. Troughs may be made from a variety of materials such as plastic insulating tape or metal foil or may be obtained commercially. Troughs are sealed with warm dental wax.

Ultra-thin sectioning. Operating procedures for commercially available ultramicrotomes differ and are described in detail in the respective

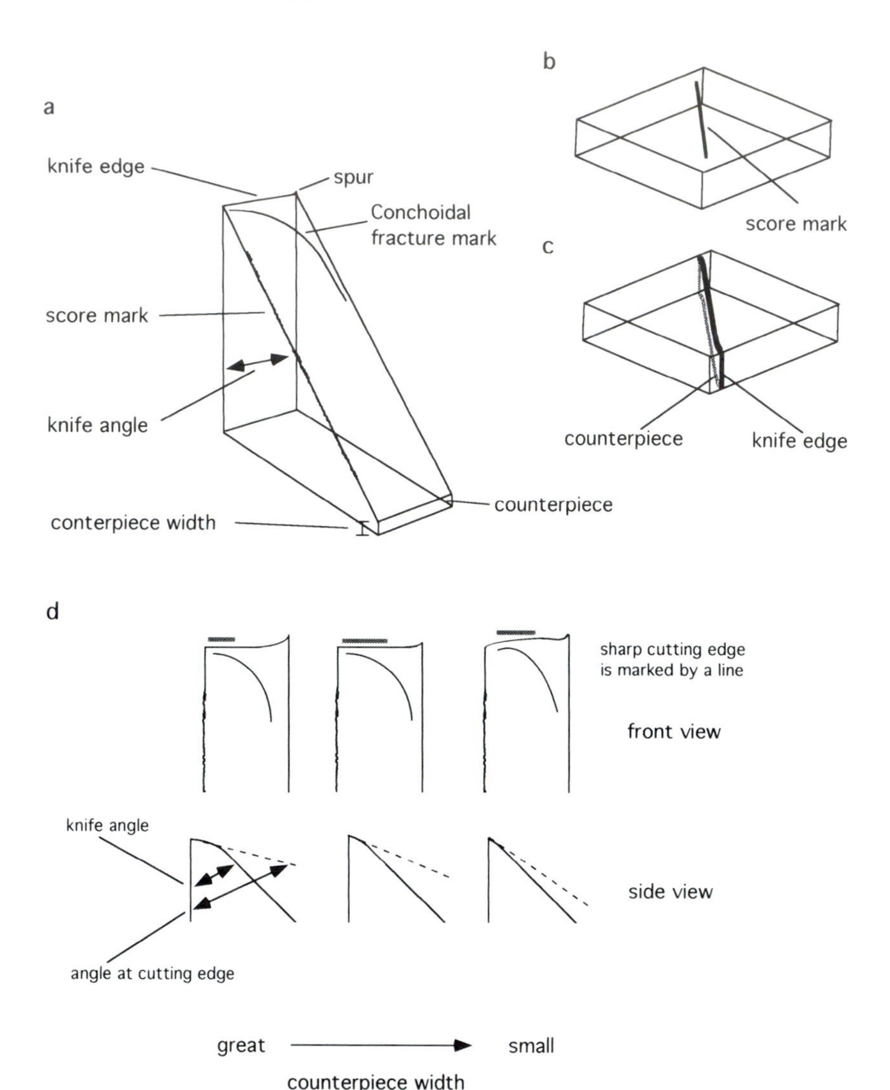

Figure 4.20. Appearance and quality of glass knives: (a) general appearance of a glass knife (without trough); (b) scored; (c) broken glass square; (d) glass knives of different qualities (see Section 4.6.3). Modified from Meek, 1976.

user's manuals. The protocols included here (*Tables 4.21* and *4.22* and *Figure 4.21*) refer to adjustment procedures common to all known standard instruments, but do not replace the manuals.

The thickness of sections floating on the water surface during microtomy can be estimated by adjusting the microtome viewing-system illumination so that the sections appear coloured on the water surface (see *Table 4.23*). The sections are picked up with a (Formvar-coated) grid (see above).

For some elemental-analysis procedures (see below), sections of minimal thickness (below 50 nm) are required. Very thin sections are best

Table 4.21. General procedure for ultra-thin sectioning

1. Mount the trimmed block
2. Mount the knife and select a clearance angle between 1–5°
3. *Perform ultra-thin sectioning:* All adjustments need to be made via observation through the microtome binocular: set the cutting edge parallel with the mesa of the resin block: this is most easily done if the mesa has a smooth, reflective surface

   - Switch on microtome lights
   - Bring the mesa face level with the cutting edge by adjusting the microtome arm height
   - Advance the knife with the coarse control, observing the reflection of its edge in the block face. The reflection aids adjustment of the knife parallel to the mesa surface and estimation of the distance between the mesa surface and the cutting edge as indicated in *Figure 4.21*
   - Move the microtome arm up and down and observe the clearance between knife and mesa carefully: if it increases or decreases, the mesa surface is not exactly vertical and the resin block has to be adjusted
   - Advance the knife with the coarse and fine controls until the clearance between knife and mesa is barely visible
   - Fill the trough with water until an almost flat meniscus causes optimum wetting of the cutting edge
   - Switch on the microtome arm automatic advance and cut some semi-thin sections in order to obtain an absolutely flat surface
   - Switch over to ultra-thin sectioning.

   Observe interference colour (see *Table 4.23*) and overall appearance of the sections: only sections with uniform colour, without scratches, wrinkles or holes are suitable for further processing
4. Move sections on the water surface with the aid of a mounted eyelash. Do not touch the sections directly during this process, but create gentle currents in the liquid to orient the sections suitable for further preparation so that they can be picked up
5. Pick up the arranged sections with a Formvar-coated specimen grid: slowly lower a grid on to the water surface, allow the grids to adsorb the sections for some seconds and then withdraw the grid
6. Drain excess of liquid off with a piece of filter paper and store the grid (sections upwards until post-staining is applied). If immunolocalization procedures are to follow, place the grid on a drop of phosphate buffer (50 mM, pH 7.0) until use (eventually refrigerated overnight)

Table 4.22. Preparation of sections below 50 nm in thickness

1. Trim the resin-embedded sample in such a way that the flat top of the pyramid (mesa) has a surface area of 0.25 mm^2 or less (it is easier to obtain sections below 50 nm with very small mesas)
2. If glass knives are used, they should be of the highest quality. Cutting semi-thin sections during knife adjustment reduces the quality of the knife edge. Ultra-thin sectioning requires readjustment of the knife edge to a new position. Cut some 80 nm thick ultra-thin sections ("silver" sections) and check section quality. If the sections are absolutely homogeneous (interference colour, no wrinkles, knife marks, etc.), proceed to step 3. If not, use a new glass knife
3. Reduce section thickness to 50 nm. If grey sections are cut, reduce feed in 1-nm steps. When sections of doubled thickness every other cut are performed, raise, step by step, the thickness until sections are performed every cut. The sections obtained by this method are of the minimal possible thickness that can be achieved with the used set-up
4. Sections in a ribbon may be stabilized and, since they are barely visible, "marked" by silver sections: after a section of the desired thickness has been cut, raise the section thickness immediately to 80 nm and cut several silver sections, then reduce section thickness again
5. For subsequent elemental analysis (see below), pick up the sections with bare grids (600–1000 mesh), i.e. grids without any support film

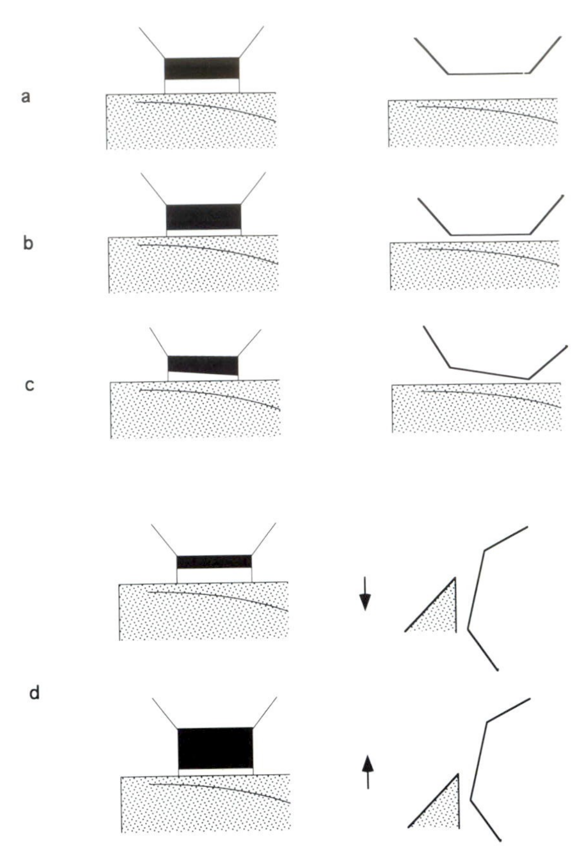

Figure 4.21. Adjustment procedure for a glass knife (*Table 4.21*). (a), (b) (left) Appearance of the reflection at the mesa surface at two different distances (right: respective top views). (c) (left) Appearance of the reflection when the mesa surface is not parallel to the cutting edge (right: top view). (d) Appearance of the reflections when the mesa surface is not exactly vertical and the microtome arm is moved up and down (see arrows; left: front view; right: side view).

Table 4.23. Determination of section thickness on the basis of interference colour

Interference colour	Section thickness	Useful for
Grey	< 60 nm	Analysis of elemental distribution [e.g. with electron spectroscopic imaging (ESI), see below]; use best quality glass knives or diamond knives
Silver	60–90 nm	"Standard" thickness for most morphological and immunocytochemical studies
Yellow–gold	90–150 nm	Easier to cut with medium quality knives, sometimes usable for immunocytochemical studies as well
Purple–blue	150–300 nm	For thick section studies (at 100 kV or with the aid of ESI; see below)
Opaque silver	> 500 nm	For high-voltage electron microscopy and light microscopy (sections are to be stained with toluidine blue)

cut with knives with very sharp cutting edges. Diamond knives provide the best results. In order to overcome the minimal thermal expansion of the instrument, the instrument illumination should be switched on 1 h prior to use; the knife holder and specimen holder should already be mounted. The mechanical feed mechanisms in modern ultramicrotomes (e.g. Reichert Ultracut series) are more precise than thermal mechanisms. During sectioning, the microtome itself should not be touched by the operator. All standard procedures to minimize transfer of mechanical vibrations have to be applied (see *Table 4.22* for subsequent preparation steps).

4.6.4 Post-staining of sections

Contrast is achieved (or enhanced, when heavy metal staining is applied during the infiltration procedure; see above) in the sectioned samples by floating the grids, with the attached sections facing downwards, on a small drop of the staining solution. Uranyl acetate heavily stains nucleic acids and also proteins, lead citrate stains membranes, nucleic acids and proteins, because lead cations bind to phosphate, carboxyl and sulphydryl groups. Osmium stains are enhanced by the application of lead citrate staining. Staining with ruthenium red solution (in combination with uranyl acetate) leads to improved contrast of wall layers (including slime and capsule), cytoplasmic membrane, nucleoid and some cell inclusions (Vogt *et al.*, 1995). Preparation of staining solutions is described in *Table 4.24*. Some examples of prepared specimens are depicted in *Figure 4.22*.

Staining with uranyl acetate and either ruthenium red or lead citrate may be combined in order to achieve maximum contrast. During lead

Table 4.24. Preparation of staining solutions for ultra-thin sections

Uranyl acetate
Suspend uranyl acetate in distilled water (up to 4%, w/v) and allow to dissolve over several hours, remove undissolved material by sedimentation in the bottle and centrifugation just prior to use. Store in a light-tight bottle at room temperature. See also *Table 4.1*

Lead citrate
a. According to Reynolds (1963):
   1. Dissolve 1.33 g of lead nitrate [$Pb(NO_3)_2$] and 1.76 g of sodium citrate [$Na_3(C_6H_5O).2H_2O$] in 80 ml of distilled water and shake vigorously for 1 min
   2. Incubate for 30 min with intermittent shaking
   3. Add 8.0 ml of 1 M NaOH and 12 ml of distilled water

   The clear solution is ready to use, faintly turbid solutions have to be centrifuged (10 min, 15000 *g*) before use; store refrigerated
b. According to Venable and Coggeshall (1965):
   Dissolve 10% (w/v) lead citrate in 0.1 M NaOH (solution in CO_2-free [boiled!] water); the turbid solution is to be centrifuged (10 min, 15000 *g*) before use

Ruthenium red (Vogt *et al.*, 1995)
Dissolve ruthenium red in distilled water or buffer solution. Combinations of ruthenium red with borate buffer (pH 9.2), cacodylate buffer (pH 7.0) and glycine buffer (pH 3.0 and pH 9.0) are described

citrate staining, crystalline precipitates of lead carbonate may occur (often along membranes or cell walls). Therefore, atmospheric CO_2 has to be excluded during the staining procedure (see *Table 4.25*). Staining with ruthenium red does not lead to precipitates.

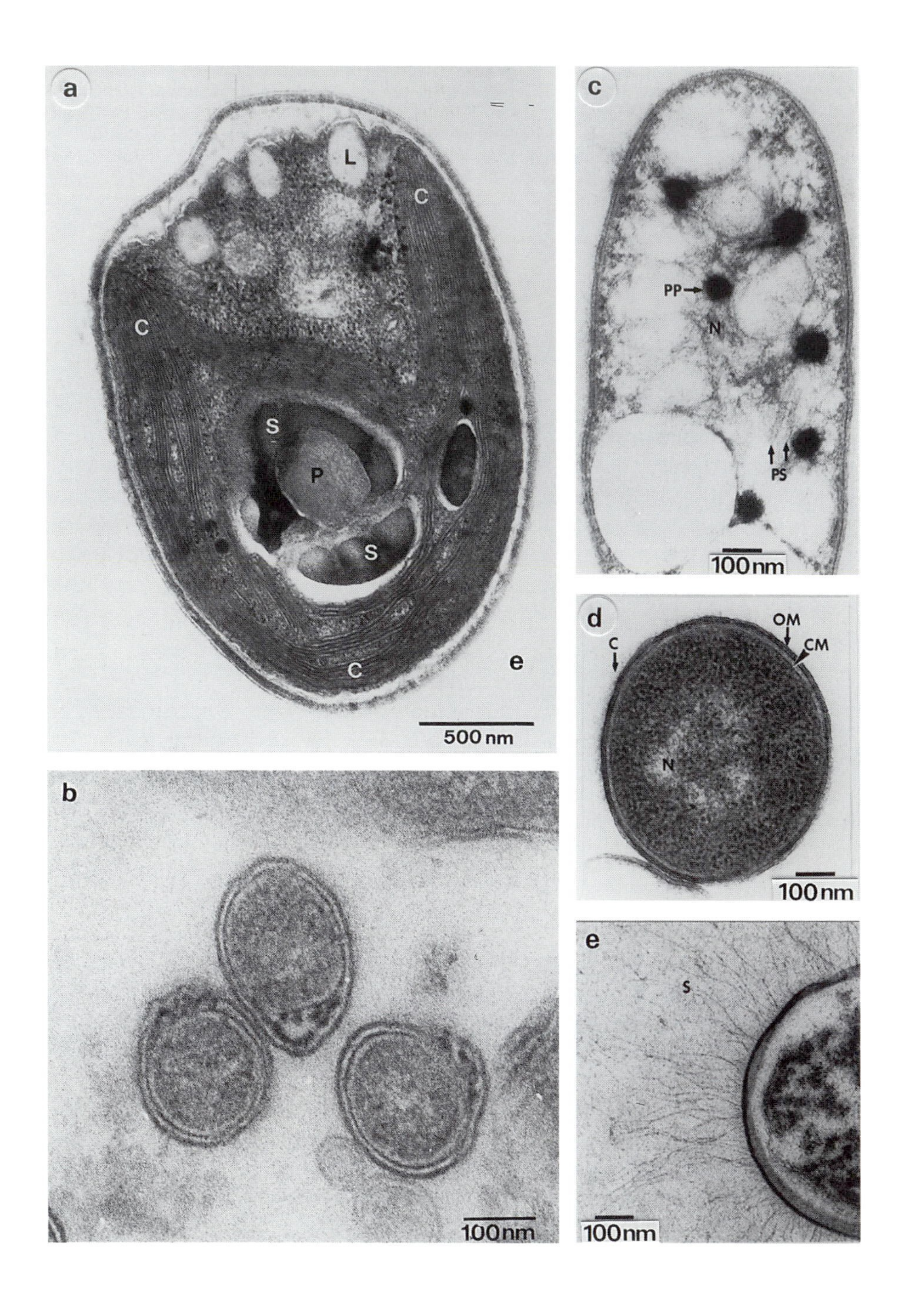

Figure 4.22. Resin-embedded ultra-thin sections. (a) The green alga *Chlorella vulgaris*; Spurr resin, post-staining with 10% (w/v) lead citrate in 0.1 M NaOH and 4% (w/v) uranyl acetate. Membraneous and non-membraneous cellular constituents are clearly visible (P: pyrenoid; S: polysaccharide sheath; C: chloroplast; L: liposomes). (b) The pathogenic spirochaete *Borrelia burgdorferi*; Lowicryl resin, post-staining with 4% (w/v) uranyl acetate. This staining solution provides sufficient contrast in specimens embedded in Lowicryl resin. Though outer and cytoplasmic membrane are clearly visible as dark contours, a differentiation between outer and inner leaflets of the respective membranes is not possible. (c) *Ralstonia eutropha* (Lowicryl resin). Staining of uranyl acetate followed by 1% (w/v) ruthenium red in cacodylate buffer, pH 7.0, provides excellent resolution of membranes (see also below, *Figure 6.2*, densitogram). (N: nucleoid; PP: polyphosphate; PS: polysheaths; see Walter-Mauruschat *et al.*, 1977). (d) *Escherichia coli*, treated as described in (c). (e) *Clostridium thermoautotrophicum* (Lowicryl resin). Staining of uranyl acetate followed by 1% (w/v) ruthenium red in glycine–KOH buffer, pH 9.0, permits visualization of slime material (s). (c)–(e) Original micrographs: R. Berker, B. Vogt; see also Vogt *et al.* (1995).

4.6.5 Cryo-ultramicrotomy

Thin sections of specimens prepared by rapid freezing methods give a more accurate picture of the overall structure of the object under study owing to superior ultrastructural preservation, because of the omission of chemical fixation and conventional dehydration methods. Though preparation of samples by freeze substitution, followed by resin embedding and conventional sectioning (see above) improves structural preservation, well-performed cryosections provide insight into the structure of specimens that have been maintained fully hydrated and uncontaminated since the initial rapid freezing procedure.

For the production of ultra-thin sections from frozen samples (see above), an ultramicrotome equipped with a cryochamber has to be used. This instrument maintains low temperatures (–130° to –50°C) throughout the sectioning procedure. Ultramicrotomes with cryochambers are available from various manufacturers (LKB, Reichert-Jung, both distributed by Reichert-Cambridge Instruments; Dupont-Sorvall). A great

Table 4.25. Staining with uranyl acetate and lead citrate or ruthenium red

1. Prepare a fresh lead citrate solution (Reynold's stain is more stable against carbonate precipitation) as described in *Table 4.24* and centrifuge a uranyl acetate solution (does not need to be freshly prepared)
2. Place 50 μl drops of uranyl acetate solution on a clean Parafilm surface; place drops of lead citrate solution on a Parafilm strip in a Petri dish containing several moistened KOH or NaOH pellets (in order to maintain a CO_2-free atmosphere)
3. Place the grids on the uranyl acetate drops with the sections facing downwards. Incubate for 3 min (e.g. for Lowicryl-embedded sections) to 15 min. If no further staining is necessary, soak away excess of liquid using a filter paper and store the grids (sections upwards). For subsequent ruthenium red staining, place the grids on drops of ruthenium red solution for up to 10 min and then blot dry. For subsequent lead citrate staining, proceed to step 4
4. Place the grids on lead citrate drops and incubate for 2–10 min
5. Wash the grids free of surplus stain by repeated dipping in a series of three 20 ml beakers of boiled (i.e. CO_2-free) double-distilled water. Alternatively, wash in a beaker containing 20 mM NaOH in double-distilled water and three beakers of double-distilled water

Table 4.26. Cryosectioning

1. Pre-cool the ultramicrotome and mount the knife. Knife and specimen holder temperatures should be of the order of –80°C to –120°C for cryoprotected and/or chemically fixed specimens (see above) or lower for specimens frozen in the native state. For samples to be subjected to X-ray microanalysis, temperatures should not be raised above –140°C. Precool all tools in liquid nitrogen or directly in the cryochamber of the ultramicrotome. An atmosphere of cold nitrogen gas in the cryochamber prevents the entrance of moisture
2. Transfer the frozen specimen (see, e.g. *Table 4.9*) to the precooled cryo-ultramicrotome. During transfer, the specimen is protected by the liquid nitrogen that remains in the cryo-cap. Most simply, "pour out" the specimen holder with liquid nitrogen in the cryochamber
3. Mount the specimen in the specimen holder, using the precooled tools
4. Trim the samples with an edge of the mounted knife or with special metal trimming edges (as, e.g. used in the Reichert-Jung FC-system). Form a flat-topped pyramid (mesa), but be aware, that, if no cryoprotection has been applied, only a thin surface layer is well vitrified: therefore, trimming a mesa would remove the best frozen parts of the specimen (freezing the specimen by the slamming technique would overcome this problem, since the technique produces a flat top, see above)
5. Adjust specimen and knife as described for conventional sectioning (see above). The sections are placed, with the aid of an eyelash, mounted on the top of, e.g. a Pasteur pipette, on the knife surface. Electrostatic interactions between the tool and the sections will lead to loss of some sections, but proper orientation is essential for the subsequent step. Use an antistatic pistol to reduce loss of sections
6. Use a drop of 2.4 M sucrose solution in water in a wire loop of about 3 mm in diameter to pick up the section placed on the knife surface (see *Figure 4.23*). Transfer the drop (with the section attached to its surface) to room temperature. The procedure has to be finished before the sucrose drop is frozen
7. Pick up the sections with a carbon-coated formvar grid. If immunolocalization is to follow (see *Table 4.37*), a carbon-coated Formvar nickel grid has to be used
8. Wash away the sucrose solution by placing the grid on the surface of distilled water. Repeat the washing procedure three times

advantage of the Reichert-Jung Cryosystem FC 4 (and later developments thereof) is the front entry system for cryotransfer devices.

Glass knives for cryosectioning should be made in knife makers according to the "balanced break" method. A glass strip (see above) is broken into two equal halves. With an equal mass of glass on each side of the score, the break is balanced and the freshly fractured surfaces are planar. By continuing to divide each piece produced into two equal halves, up to 16 squares can be made from a glass strip 40 cm in length. All the squares produced have straight sides and precise right-angled corners. The glass knives should have a real knife angle closest to 45° and be of the best quality for maximum sharpness (Griffiths *et al.*, 1983). Knives should be prepared shortly before use. The stability of the cutting edge may be improved by a light tungsten coating (Roberts, 1975). The knives are normally used without a trough and are kept dry. The specimen is transferred, in liquid nitrogen, to the precooled specimen holder. Sectioning is performed, in principle, according to the conventional procedure under subsequent cooling. The ultra-thin sections are picked up by adsorption on to the surface of a drop of sucrose in a small metal loop. During this transfer the section is warmed up to room temperature and transferred to a (carbon-coated) Formvar grid.

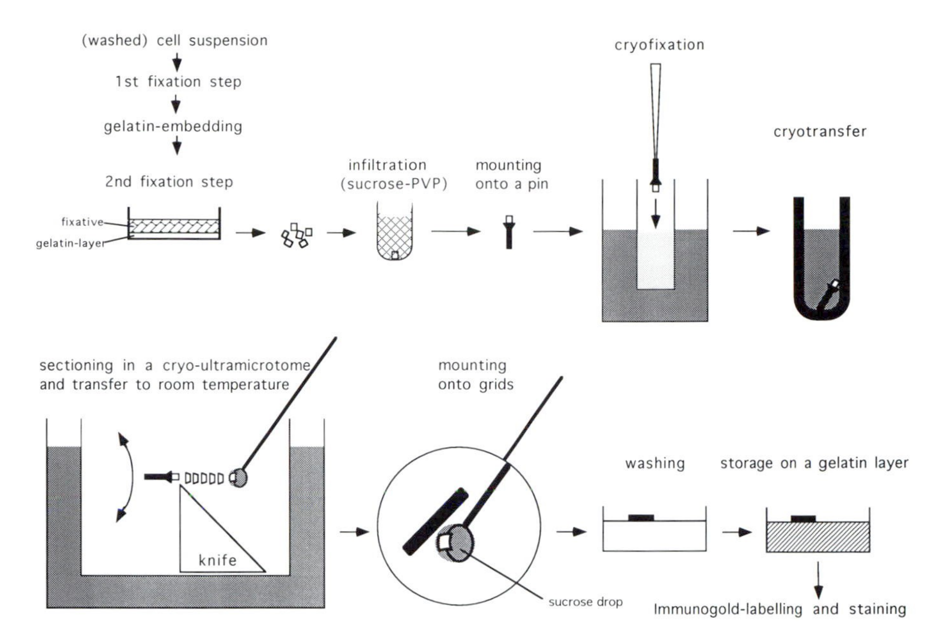

Figure 4.23. Most important preparation steps for cryofixation and subsequent cryosectioning procedures.

A detailed outline for a cryosectioning procedure is given in *Table 4.26*. In *Figure 4.23* the whole preparation procedure (including cryofixation; see above) is summarized. The preparation method is suitable for subsequent morphological studies of the sections and immunolocalization procedures.

For subsequent immunolocalization, place the grids on ice-cooled gelatin (2%, w/v, aqueous solution). The grids may be stored on the gelatin surface overnight at 4°C. Alternatively, the sections may be stained as described in *Table 4.27*.

Table 4.27. Staining of cryosections

1. Prepare a 2% (w/v) methylcellulose solution in water and centrifuge for 90 min at 250000 *g* (the solution may be stored refrigerated for several weeks). For preparation of the staining solution, add the desired amount of a uranyl acetate stock solution (4% w/v). A final concentration of 0.3% (w/v) uranyl acetate is suitable for the first trial. Optimize the uranyl acetate concentration by checking the contrast of the specimen in the electron microscope
2. Place grids on to two drops of the staining solution for several seconds each and then on to a third drop of staining solution for 3 min
3. Pick up the grids with a wire loop and soak away excess of staining solution. Allow the grids to dry for 20 min

   For subsequent elemental analysis, wetting of the sections is not recommended. Once sectioned, the specimens are further processed by mounting, by means of pressing the sections in a specially designed device, on to a grid surface, (Tvedt *et al.*, 1984) and then subjected to subsequent freeze drying (Geymayer *et al.*, 1977)

Table 4.28. Techniques for the localization of biological macromolecules by electron microscopy

Problem	Specimen treatment	Marker system (technique; references)
Localization of specific proteins (antibody–gold, see *Figure 4.24a* and *b* or carbohydrate complexes (lectin–gold) inside cells (*Figure 4.25*)	Ultra-thin sections of resin-embedded samples (Lowicryl, LR gold) subjected to immunolocalization (*Table 4.33*)	Antibody–colloidal gold; lectin–colloidal gold; post-embedding technique (Roth *et al.*, 1980; Horisberger, 1985)
Localization of specific proteins on the cell surface or subcellular particles (modified technique useful for SEM studies)	Incubation of cells or subcellular fractions with the marker system prior to conventional embedding in, e.g. Spurr resin (*Table 4.35*)	Antibody–colloidal gold; antibody–ferritin; lectin–colloidal gold; pre-embedding technique (Rohde *et al.*, 1988)
Localization of specific proteins on the surface of, e.g. membrane vesicles (see *Figure 4.24c* and *d*)	Subcellular fraction adsorbed on to a carbon-coated Formvar-grid subjected to immunolocalization (*Table 4.36*)	Antibody–colloidal gold; antibody–ferritin; lectin–colloidal gold; whole-mount technique (Acker, 1988)
Detection of viral surface antigens	Aggregation of virus particles by incubation with specific antibodies (e.g. specific to surface epitopes) followed by negative staining	Virus particle and a specific antibody (no additional markers) (Miller and Howell, 1997)
Localization of single epitopes (subunits) and their relative position to each other on the surface of the object (*Figure 4.26*)	Homogeneous protein preparation (especially useful for large proteins or viruses) subjected to localization of epitopes (see Section 4.7.4)	Antibody, Fab-fragment; epitope labelling (Hermann *et al.*, 1991)
Localization of biotin-containing enzymes, localization of the biotin site	Subcellular fraction, homogeneous enzyme preparation subjected to localization	Avidin–gold (Mayer and Rohde, 1988); avidin (Däkena *et al.*, 1988)
Detection of enzyme activities	Incubation of cells or subcellular fractions with the marker system prior to conventional embedding in, e.g. Spurr resin	Specific enzyme substrate coupled with precipitation of electron-dense marker (Hayat, 1973; Wohlrab and Gossrau, 1992)
Detection of macromolecular components acting as enzyme substrates (e.g. starch, nucleic acids)	Labelling of enzyme substrates with gold particles coated with the relevant enzmes in ultra-thin sections of samples	Enzyme–colloidal gold (Bendayan, 1985)
Localization of metabolites in the cell	Incubation of metabolically active cells with the precursor, embedding in	Metabolic precursors (e.g. sugars or amino acids) labelled with

(Continued)

	conventional resin or by application of cryotechniques, treatment with a specific photoemulsion	radioactive markers (autoradiography; Gregg and Reznik-Schüller, 1984)
Localization of specific DNA or RNA (preferably mRNA) sequences in cells	Hybridization DNA/RNA of fixed samples with the marker system followed by embedding and sectioning. Detection of the marker system by autoradiography or with streptavidin–colloidal gold	RNA or DNA oligonucleotide labelled with different markers (*in situ* hybridization; Egger *et al.*, 1994)

4.7 Localization of macromolecules

Single epitopes of enzymes in pure preparations can be localized by epitope mapping with the aid of specific antibodies. The localization of proteins in organisms is performed by application of labelling procedures based on specific binding of colloidal gold or other electron-dense markers. A survey of the techniques used is given in *Table 4.28*.

4.7.1 Preparation of marker systems for localization procedures

Immunocytochemical localization, that is, labelling with the aid of a specific (monoclonal or polyclonal) antibody and colloidal gold (5–20 nm), as electron-dense marker system, is the most widely used localization method. The most common immunocytochemical localization procedures use colloidal gold particles coupled with proteins that bind to the specific (primary) antibody. These proteins are either Protein A or a secondary antibody directed against the Fc-fragment of the specific antibody. Preparation of the marker systems is performed according to established procedures (Rohde *et al.*, 1988; Roth *et al.*, 1978); the markers are also available from various commercial sources.

Monodisperse colloidal gold particles may be prepared in sizes ranging from 2 to 150 nm (Frens, 1973). The size of the gold particles depends mainly on the strength and quantity of the reducing agent used. Phosphorus, ascorbic acid, sodium citrate and tannic acid are widely used reducing agents. To date, a wide range of proteins have been successfully coupled with colloidal gold. Geoghegan and Ackerman (1977) found that adsorption of proteins to colloidal gold depends on the pH of the solution and the isoelectric point of the protein that is to be coupled. Adsorption is also influenced by the interfacial tension and the solubility of the molecule (see *Tables 4.29*, *4.30* and *4.31* for preparation of colloidal gold and the coupling procedure).

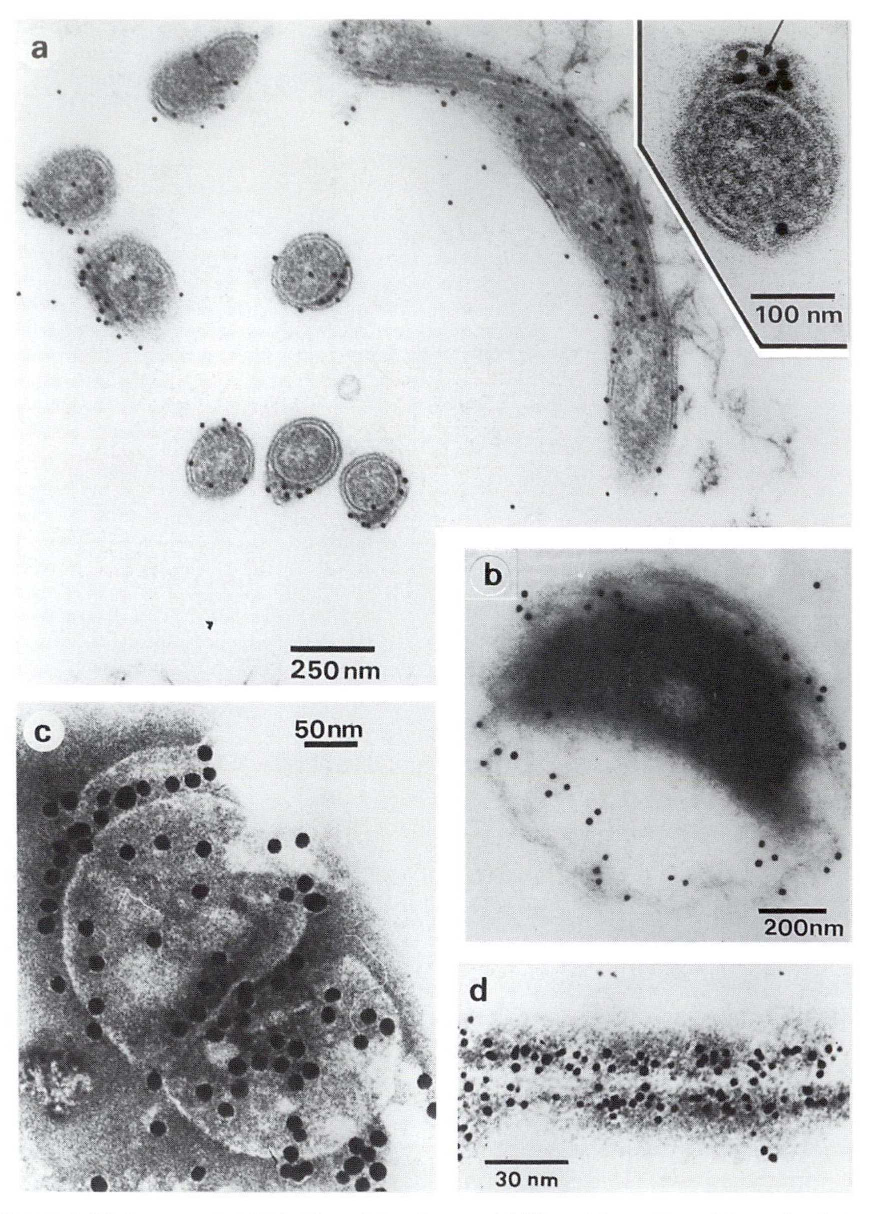

Figure 4.24. Immunogold labelling of specimens. (a) Ultra-thin section of the spirochaete *Borrelia burgdorferi*, embedded in Lowicryl resin, labelled with an antibody directed against an endoflagellar component (site of the endoflagella is marked in the inset). Note the variations in appearance of the cells and distribution of gold markers depending on the section plane (small, approximately round cross-sections and a large longitudinal section). See also Eiffert *et al.* (1992). (b) A plasmolysed cell of *Ralstonia eutropha* reveals that the antigen (the "membrane-bound" hydrogenase) is located in the (artificially widened) periplasm (see Eismann *et al.*, 1995). (c) Cytoplasmic membrane vesicle from *Methanobacterium thermoautotrophicum*, labelled with an antibody directed against a membrane spanning protein (see Stupperich *et al.*, 1993). Preparation according to the "whole-mount" procedure (see *Table 4.36*). (d) Isolated endoflagellum (see descriptions in (b), obtained by ultracentrifugation of the cellular constituents, prepared according to the "whole-mount" procedure (*Table 4.36*; see also Eiffert *et al.*, 1992.).

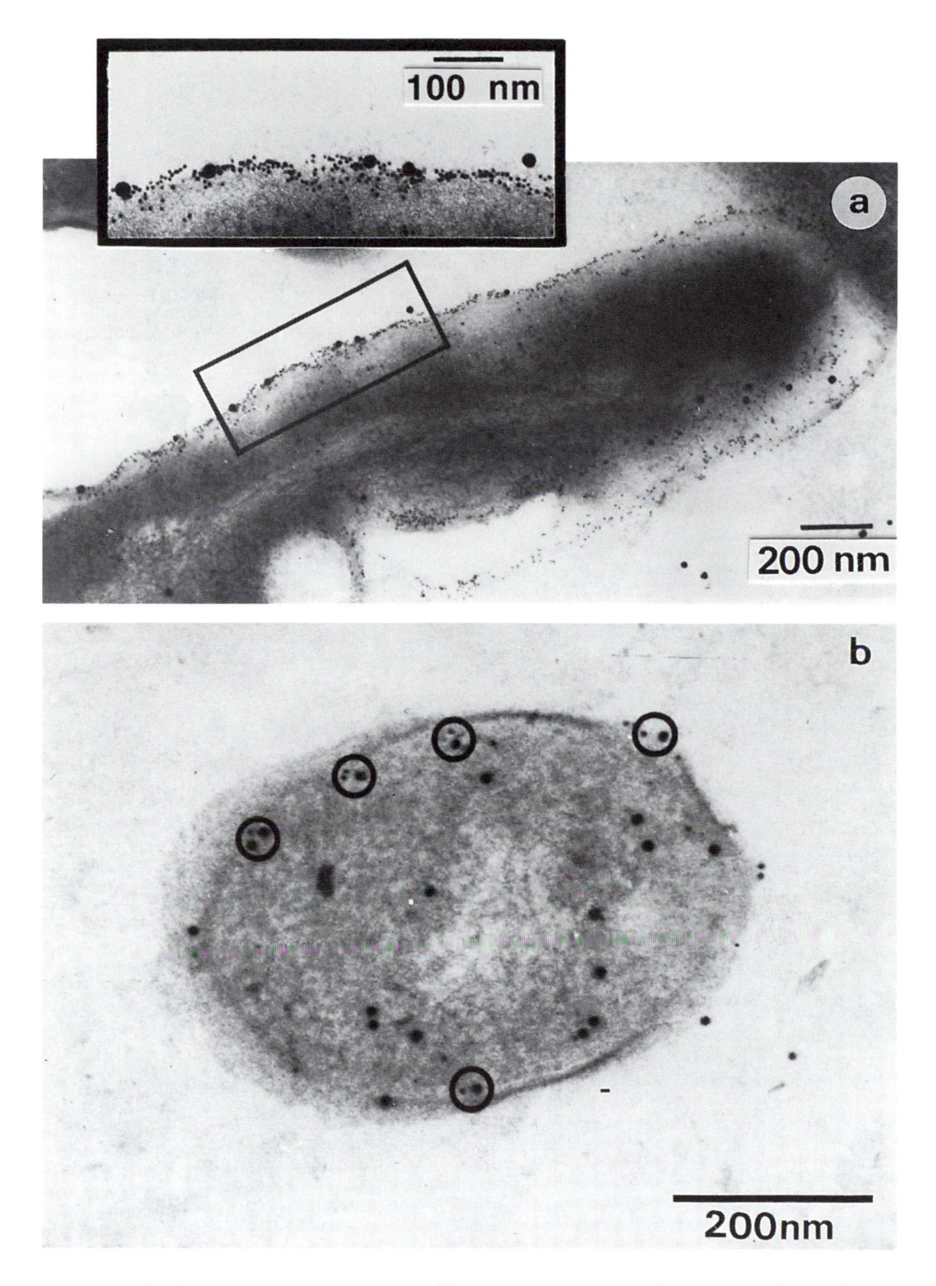

Figure 4.25. Immunogold double-labelling procedures. (a) Bacterial cell (not further characterized) double labelled with lectin from *Arachis hypogaea* coupled with gold (3–5 nm) and Concanavalin A coupled with gold (15–20 nm). Binding sites are located at the cell wall (original micrograph: M. Kämper). (b) Co-localization of two enzymes as parts of a membrane-bound complex (H_2-heterodisulphide oxidoreductase complex) in *Methanobacterium thermoautotrophicum*. Double labelling with antibodies directed against two enzymes of the complex reveal that the enzymes are in close vicinity to each other when located at the cell periphery (marked by circles; small gold particles of 6 nm diameter and large particles of 10 nm diameter). One enzyme is also present in the cytoplasm (large gold particles of 10 nm diameter).

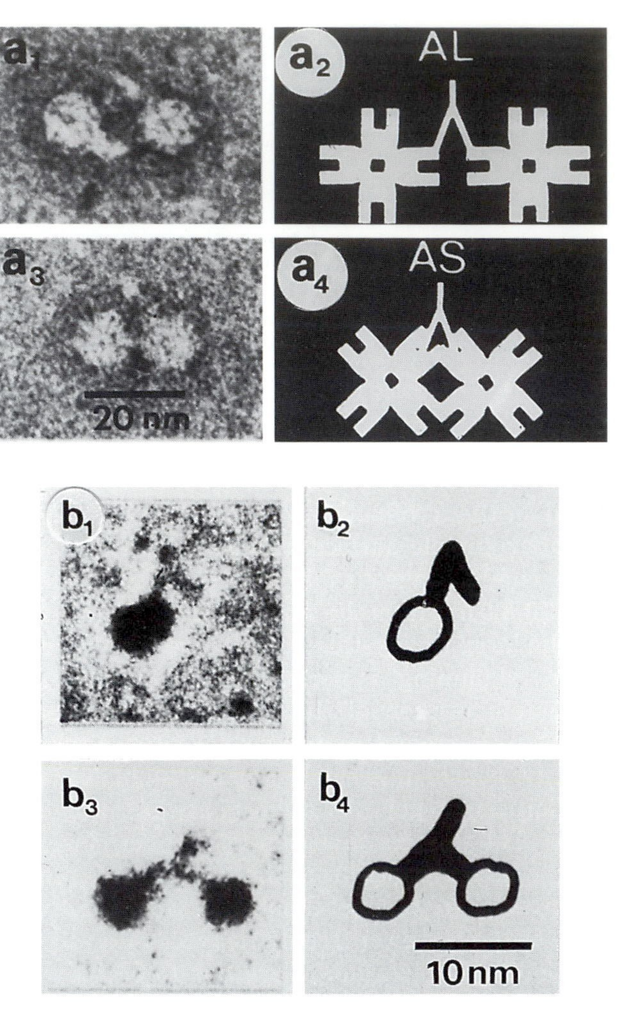

Figure 4.26. Epitope mapping of enzymes with IgG antibodies. (a) Localization of the large and small subunit of the hexadecameric (L_8S_8) D-ribulose-1,5-bisphosphate carboxylase/oxygenase (RuBisCO) from *Ralstonia eutropha* (68). Purified RuBisCO was labelled with polyclonal antibodies directed against the large enzyme subunit (AL, see a_1 and a_2) and the small subunit (AS, see a_3 and a_4). Antibody cross-linked enzyme molecules (a_1, a_3) show different intermolecular distances depending on the antibody used, leading to a model of the subunit topology of the enzyme (a_2, a_4). See also Holzenburg and Mayer (1989). (b) Imaging of Y-shaped monoclonal antibodies bound to the Fo-complex of the *Escherichia coli* ATP synthase. b_1, b_2: Antibody bound to one molecule. b_3, b_4: Antibody cross-linking two molecules. See Birkenhäger *et al.* (1995).

Since colloidal gold is the most widely used marker system for localization, it is available from numerous suppliers, both free and coupled with various secondary reagents (antibodies, biotin, lectins, etc.). Gold lots from various suppliers show wide variations in quality. Monodisperse colloidal gold of 5–15 nm has a dark red to wine-red colour. Even a slight violet or blue–violet coloration indicates clustering of the gold particles. Clustered particles are useless for electron-microscopic

Table 4.29. Preparation of colloidal gold (approximately 10 nm particle size)

1. Siliconize the inner surfaces of all vials and flasks to be used for the preparation and storage of colloidal gold by rinsing with a solution of Sigmacote™
2. Dry the glassware and microcentrifuge tubes for 2 h at 110–130°C
3. Mix 4 ml of 1% (w/v) sodium citrate solution and 16 ml of double-distilled water in a 50 ml Erlenmeyer flask
4. Mix 1 ml of 1% (w/v) tetrachloroauric acid solution and 79 ml of double-distilled water in a 300 ml Erlenmeyer flask
5. Incubate both solutions for 15–20 min in a water bath (60°C)
6. Combine both solutions under immediate and gentle rotation of the flask. Avoid excessive magnetic stirring! A violet colour indicates the reduction of tetrachloroauric acid. The process is complete after 60 min
7. Boil the solution for 5 min; this results in a deep red colour
8. Store the solution refrigerated in a siliconized glass vessel (stable for several weeks; check by electron microscopy before use)

Table 4.30. Preparation of colloidal gold (approximately 5 nm particle size)

Use siliconized glassware (see *Table 4.29*, steps 1 and 2)

1. Mix 4 ml of 1% (w/v) sodium citate solution, 1 ml of 1% tannic acid solution, 1 ml of 25 mM potassium carbonate and 14 ml of water
2. Mix 1 ml of 1% (w/v) tetrachloroauric acid solution and 79 ml of double-distilled water in a 300 ml Erlenmeyer flask
3. Combine both solutions under immediate and gentle shaking of the flask. The solution turns red immediately
4. Boil the solution for 5 min; this results in a slight deepening of the red colour
5. Add hydrogen peroxide solution to a final concentration of 0.3–0.5% (v/v) to remove residual tannic acid
6. Incubate overnight at room temperature; then incubate for 30 min at 90°C to remove hydrogen peroxide
7. Store the solution refrigerated in a siliconized glass vessel (stable for several weeks; check by electron microscopy before use)

applications (although they may still function in light-microscopic histochemical or biochemical tests).

Colloidal gold complexes may be prepared with a minimum of laboratory equipment. The quality of the preparations (particle size distribution, clustering) has to be carefully checked (by electron microscopy) before they are coupled with proteins, or other reagents, and before use. In double-labelling experiments, the use of two differently sized gold particle preparations, coupled with secondary reagents, permits differentiation between the locations of two defined binding sites.

During the immunolocalization procedure, the specimen is incubated with the primary antibody (preferably IgG) and this is then detected by means of the marker system via binding of Protein A or the secondary antibody. Most bacterial protein types may be detected by this method if the concentration of epitopes is high enough (at least 100 per bacterial cell). Labelling of ultra-thin sections ("post-embedding labelling") allows the assignment of epitopes to the cell periphery (either cytoplasmic membrane or periplasm), storage granules, the nucleoid or the "free" cytoplasm. It remains difficult to resolve whether an epitope is, for example, bound to the cytoplasmic membrane or located in the periplasm. The

Table 4.31. Determination of the optimum protein concentration and coupling procedure to yield colloidal gold conjugates

1. Adjust pH of the colloidal gold solution (as a general rule 0.5 pH unit above isoelectric point of the protein; for Protein A, pH 5.5–6, for IgG pH 9.0 is recommended). Use 0.1 M HCl or 0.2 M K_2CO_3
   Dialyse the protein solution against an appropriate buffer of low ionic strength (for Protein A, simple dilution in distilled water is sufficient; for IgG, dialysis against 2 mM borax, pH 9.0, is recommended)
2. Mix aliquots (250 μl) of colloidal gold solution with 50 μl aliquots of various protein dilutions (1:8, 1:16, 1:32 ... 1:1024; use siliconized microcentrifuge tubes). Incubate for 10 min at room temperature
3. Add 30 μl of a 10% (w/v) sodium chloride solution to each protein dilution step. Incubate for 10 min at room temperature; a colour change from red to blue (aggregation of colloidal gold particles) can be observed in vials where colloidal gold is insufficiently stabilized by the protein dilution. Sufficiently stabilized samples retain the bordeaux-red colour
4. For preparation of the bulk amount of protein–gold complexes, add protein solution in 10% excess of the minimal stabilizing concentration, determined from the stability test (steps 2 and 3). Incubate for 10 min at room temperature
5. Check the stability of the solution by adding 10% (w/v) sodium chloride solution to a small sample as described in step 3. The sample should retain its wine-red colour (if so, continue with step 6). If it does not, the protein concentration is too low; check the results of the optimal protein concentration determination (steps 2 and 3)
6. Add 5 ml of 1% (w/v) polyethyleneglycol solution (PEG 20 000; recommended for Protein A and other proteins) or add 7.5 ml of 8% (w/v) BSA (recommended for IgG; adjust to pH 9.0) to 50 ml of the protein–gold solution. Incubate for 5 min at room temperature
7. Centrifuge for 30 min at 30000 *g* (10–15 nm particle size) or for 45 min at 85000 *g* (5 nm particle size) and 4°C. Centrifugation results in formation of a dark red sediment consisting of fully stabilized protein–gold complexes. Particle aggregates or incompletely stabilized gold particles precipitate on to the centrifuge tube wall, forming a dark spot. See *Figure 4.27*
8. Do not decant the supernatant as this disturbs the sediment! Remove the sediment carefully using a pipette and then dilute it in 10 ml of potassium phosphate buffer, 50 mM (pH 6.9), containing 0.9% (w/v) NaCl and 0.02% (w/v) PEG 20000 (recommended for Protein A and other proteins) or dilute it in 20 mM Tris-HCl buffer, containing 0.5% (w/v) NaCl, pH 8.5, containing 1% (w/v) BSA (recommended for IgG)
9. Recentrifuge the resuspended sediment under the same conditions (step 7). Suck the sediment off and store at 4°C (do not freeze!). Add some sodium azide crystals as preservative

resolution of this technique is no greater than 15 nm, as the colloidal gold particle is held at a distance of 10–15 nm from the epitope due to the space occupied by primary (e.g. an antibody) and secondary (e.g. a Protein A shell) reagents. In order to decide whether a protein is located in the periplasm or attached to the cytoplasmic membrane, the periplasmic space may be artificially widened by plasmolysis of the cells prior to fixation and embedding (Eismann *et al.*, 1995; *Figure 4.24b*). After post-embedding labelling, the distribution of markers is now easier to interpret. To confirm a specific localization on the surface of a compartment (e.g whether an epitope is exposed to the cytoplasmic or the periplasmic side of the cytoplasmic membrane), preparation of protoplasts and inside-out membrane vesicles may be useful. The samples are then subjected to the labelling procedure prior to fixation and embedding (pre-embedding labelling, see, e.g. Rohde *et al.*, 1988). Detailed pre-embedding protocols are given below.

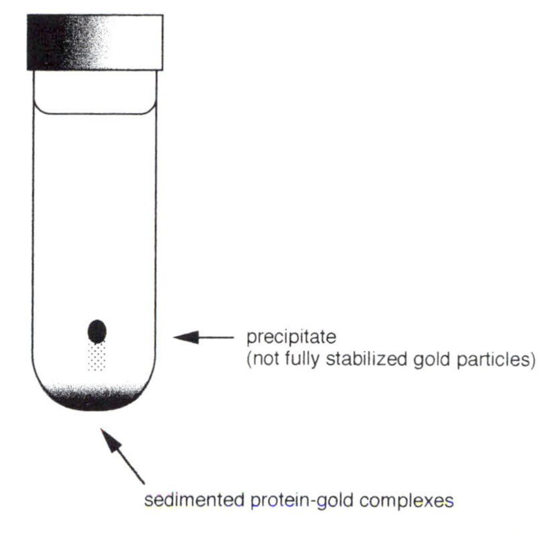

Figure 4.27. Appearance of protein-stabilized colloidal gold (see *Table 4.31*, step 6) after centrifugation. The sedimented particles at the bottom of the centrifuge tube are fully stabilized.

Several new developments potentially increase the application range and sensitivity of the method. Dodecaborane clusters may be coupled via polylysine dendrimers with antibodies or Fab fragments. The boronated antibodies have a detection limit lower than that of conventional gold conjugates in immunolocalization experiments. Detection of boron in immunolabelled sections (less than 50 nm thick) is performed via element specific imaging in energy-filtering TEMs (see below). The experimental outline is described by Qualmann *et al.* (1996) and Kessels *et al.* (1996). Boronated Protein A, instead of colloidal gold, was used by Bendayan *et al.* (1989) for the localization of various antigens in tissue sections and isolated cells.

If purified protein is available for the production of either monoclonal or polyclonal antibodies, immunolocalization may be performed using fusion tags or antibodies directed against these protein epitopes. The green fluorescent protein (GFP) of *Aequorea victoria* is a unique *in vivo* reporter for monitoring dynamic processes in cells or organisms (Gerdes and Kaether, 1996). As a fusion tag GFP can be used to localize proteins and to follow the dynamics of the compartments to which the proteins are targeted at the light-microscopical level. GFP-tagged proteins may be located in TEM preparations on formation of photo-oxidized, electron-dense diaminobenzidine (DAB) precipitates at the sites where GFP is present (Monosov *et al.*, 1996). The GFP may also be detected with (commercially available) anti-GFP antibodies.

If the sequence of a section of the protein is known, epitope-specific antibodies may be produced. Several software packages (e.g. Wisconsin GCG Sequence analysis software package, see Devereux *et al.*, 1984) allow the determination of possible epitopes in a given amino acid sequence (derived from the DNA sequence). The epitope (10–15 amino acids) is synthesized, and the resulting peptide is coupled with a carrier molecule and used for immunization.

The use of gold-coupled lectins has broadened the range of applications of immunocytochemical localization (*Figure 4.25a*). Coupling of lectins with colloidal gold is performed as described below, but pH adjustment has been described as essential for proper complex formation (Horisberger, 1985). In practice, the yield of usable gold conjugates depends on the lectin species used. The lectins Bandeiraea (from *Bandeiraea simplicifolia*) and Concanavalin A (from *Canavalia ensiformis*) are very suitable for colloidal gold stabilization. Others, such as Arachis (from *Arachis hypogaea*) and Lentil (from *Lens culinaris*), are less useful. Gold particles tend to precipitate after the second centrifugation step during the coupling procedure (see step 8 in *Table 4.31*). Wheat-germ agglutinin (from *Triticum vulgaris*) is the most difficult to handle and must be cross-linked to bovine serum albumin (BSA) prior to gold coupling (Horisberger, 1985).

The electron-microscopic localization of enzyme activity in cells ("catalytic histochemistry" or "electron-microscopic cytochemistry") allows the direct detection of catalytic activities. This method has mainly been applied to tissue sections (on the light-microscopic level) in biomedical research (comprehensive surveys on preparation procedures for electron microscopy are provided by Hayat, 1973; and Wohlrab and Gossrau, 1992). Most of the detection procedures are applicable to electron-microscopic preparations, because the catalytic reactions may be coupled with precipitation of electron-dense agents. In bacterial cells, the method has been applied to the localization of ATPase and cytochrome oxidase (Tauschel, 1988) in order to elucidate the involvement of these enzymes in flagellar motor bioenergetics. A detailed experimental procedure is described in Tauschel (1988).

Preparation of colloidal gold. *Tables 4.29* and *4.30* describe the procedure for preparation of colloidal gold of 5 and 10 nm in particle size with narrow size range distribution. These preparations are useful for double-labelling methods. The subsequent coupling of protein molecules is described in *Table 4.31* and mainly refers to coupling of Protein A and IgG, but may be also be adapted to coupling of other proteins (see also Beesly, 1989).

Preparation of wheat-germ agglutinin/BSA. Cross-linking may be performed successfully using a solution containing 2 mg ml^{-1} wheat-germ agglutinin and 2 mg ml^{-1} BSA in 10 mM potassium phosphate buffer at pH 7.0. After addition of glutardialdehyde (0.25% v/v stock solution) to a final concentration of 0.025%, the mixture is kept at room temperature for 1 h with gentle shaking. Any precipitate is then removed by filtration (0.2 μm pore size sterile filter).

Table 4.32. Removal of antibodies that bind to fungal cell walls

1. Rinse fungal mycelium (5 ml of packed cells) extensively with an appropriate buffer (1 l of, e.g. 50 mM PBS, pH 7.0) in a Büchner funnel
2. Fill a small chromatography column with the washed mycelium. Load the column with up to 500 μl of antiserum. Allow the antiserum to incubate in the mycelium for 2 h at room temperature
3. Wash the column with several millimetres of PBS. The antiserum elutes immediately, but cell-wall specific antibodies are retained
4. Depending on the properties of the mycelium, the antibodies may be contaminated with extracellular fungal components. In order to maintain long-term stability, the antiserum may be further purified by affinity chromatography with Protein A–sepharose

Pretreatment of primary antibodies. The main requirement for immunolabelling methods is high antibody specificity. It is recommended that antisera are purified by affinity chromatography using Protein A–sepharose (for rabbit antisera) or Protein G–sepharose (for mouse antisera) materials. Use of whole antisera for immunolocalization is possible, but may result in non-specific background labelling. Purification of antisera by affinity chromatography has the advantage that only intact antibodies (or Fc fragments) of the IgG subspecies remain in the resulting antibody solution. Thus, no disturbing antibody species or Fab fragments, which may react with the antigen but not with the marker, are present. This purification method is best used in conjunction with subsequent utilization of a Protein A-labelled marker.

Polyclonal antisera may contain antibodies directed against "common antigens" in the cell walls of Gram-negative bacteria and yeasts, because some pathogenic members of these groups may cause infections during immunization. This often results in the labelling of bacterial and especially fungal cell walls supposedly directed against quite different antigens. Removal of fungal cell wall antibodies may be achieved by one-step chromatography of the antisera on washed fungal mycelium (e.g. mycelium from the fungus that is subjected to the immunolocalization procedure; *Table 4.32*).

4.7.2 Localization procedures

The immunocytochemical localization of antigens in ultra-thin sections of cells is performed as described in *Table 4.33* (*Figures 4.24* and *4.25*) (Hoppert and Mayer, 1995; see also the scheme presented in *Figure 4.28*). The concomitant localization of two antigens helps reveal possible interactions and has been, for example, successfully applied for the elucidation of structure–function relationships in multi-enzyme complexes and metabolic pathways in bacteria (see Freudenberg *et al.*, 1989 and Aldrich *et al.*, 1992). The procedure is described in *Table 4.34*.

Labelling with lectin–gold complexes is performed as in *Table 4.33* but with some modifications: step 2 may be omitted, and in step 4 the enzyme-specific IgG antibody is replaced by the colloidal gold. The pH value of the incubation buffer should be slightly above the isoelectric point of the coupled lectin. Steps 5–8 are omitted.

Table 4.33. Outline immunolocalization procedure for ultra-thin sections

1. Immediately after the freshly cut sections are picked up, place the grids, with the sections facing downwards, on drops of 50 mM PBS, pH 7.0. This step is performed to prevent loss of reactivity of, e.g. antigens due to drying of the sections
2. Place the grids on drops of 1% (v/v) hydrogen peroxide for 5 min. This etching procedure results in a better exposition of antigens on the surface of the section. Rinse the grids extensively in distilled water
3. Place the grids on drops of 0.5% (w/v) skim milk solution for 5 min. This is done to reduce non-specific reactions of antibodies with the section[a]
4. Transfer the grids on to drops of PBS containing various concentrations of enzyme-specific IgG antibodies (incubation time: 2 h). A first experiment may be performed with a 1/10, 1/100, 1/1000 (...) dilution series. Evaporation of water is minimized by covering the samples
5. Rinse the grids extensively ("jet wash", i.e. running a stream of buffer from a washing bottle across the grids held with forceps), for 20 sec, in PBS containing 0.05% (v/v) Tween 20
6. Incubate twice, for 2 min on drops of PBS containing 0.05% (w/v) Tween and then for 2 min on a drop of PBS without Tween
7. Repeat step 3
8. Incubate on drops of the appropriate Protein A (or secondary antibody) coupled gold dilution for 30 min, prepared in PBS containing 0.5% (w/v) BSA. A first experiment may be performed with a 1/10, 1/100, 1/1000 dilution series. Dilutions around 1/30 (10–15 nm gold particles) and 1/80 (5-nm particles) are usually found appropriate. Note: the PAG solution has to be centrifuged for 3 min at ~20000 *g* and room temperature to remove larger particle clusters immediately prior to use
9. Remove the grids from the surface of the drops and rinse extensively ("jet wash") with PBS solution for 20 sec
10. Incubate twice for 2 min on drops of PBS at room temperature
11. Transfer the grids onto drops of double-distilled water and incubate for 1 min at room temperature. This washing step removes any residual buffer salts that may cause precipitates during staining. The step should be omitted if the antibody–antigen reaction is weakened (seen in some monoclonal antibody species)
12. Dry the grids by blotting on filter paper, but prevent direct contact between the grid surface and the surface of the filter paper (this is done as depicted in *Figure 4.3*, step 7)
13. Transfer the grids onto drops of 4% (w/v) uranyl acetate solution. Incubate for 3 min at room temperature, then repeat step 12

[a]In addition, or as an alternative to step 3, non-specific antibody–antigen and especially sample–Protein A-gold (PAG) interactions can be quenched by the addition of 0.1 to 2.0% (w/v) casein hydolysate to the antibody and/or PAG solution prior to use.

The "pre-embedding" procedure (Acker, 1988; Hoppert and Mayer, 1995) allows enzymes on the surfaces of cells and vesicles to be localized and is mainly used to detect cell surface antigens and extracellular enzymes attached to the cell surface (Rohde *et al.*, 1988, *Table 4.35*). Smaller isolated cellular constituents (membrane vesicles, cell appendages and subcellular components; *Figure 4.24c* and *d*) may be labelled by application of the whole-mount procedure (*Table 4.36*).

Immunogold labelling of cryosections. The labelling of cryosections (*Figure 4.29*) may, in principle, be carried out according to the corresponding conventional procedure (see also Griffiths *et al.*, 1984). Washing steps (in PBS containing 0.02%, w/v, potassium chloride) performed before and after the incubation with the specific antibody and the colloidal gold solution have to be performed very gently, that is, by placing

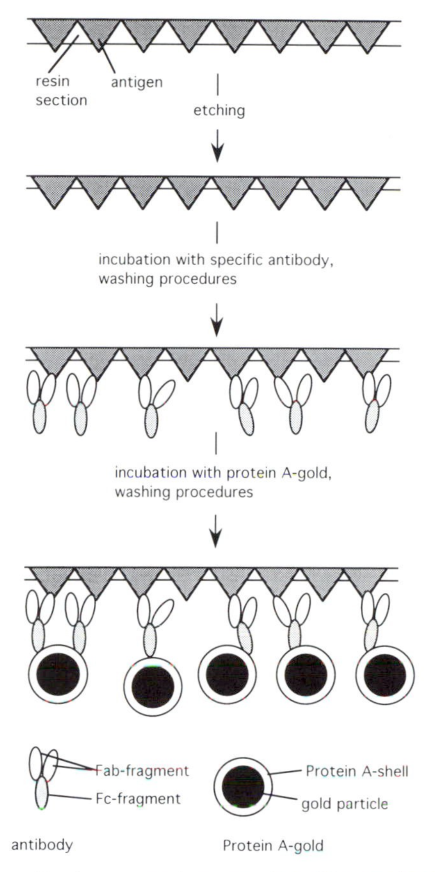

Figure 4.28. Immunolocalization procedure on ultra-thin sections. Schematic drawing illustrating the principle of the localization of antigen (see *Table 4.33*) with specific antibodies and colloidal gold (Protein A–gold) at molecular dimensions.

Table 4.34. Double-labelling procedure

1. Perform the immunolocalization with the first antigen as described in *Table 4.33*, steps 1–10. In step 8, use PAG of defined size (e.g. 10 ± 2 nm diameter)
2. Incubate the grids on drops of Protein A in 50 mM PBS, pH 7.0 (0.25 mg ml^{-1} of Protein A) for 30 min in order to saturate any remaining Protein A-binding sites
3. Remove the grids from the surface of the drops and rinse extensively ("jet wash") with PBS solution for 20 sec
4. Incubate twice, for 2 min, on drops of PBS
5. Transfer the grids on to drops of PBS containing the second specific IgG antibody (incubation time: 2 h). Evaporation of water is minimized by covering the samples
6. Rinse the grids extensively ("jet wash"), for 20 sec, in PBS containing 0.05% (v/v) Tween 20
7. Incubate twice, for 2 min, on drops of PBS containing 0.05% (w/v) Tween and then for 2 min on a drop of PBS solution without Tween
8. Incubate for 30 min on drops of the appropriate PAG dilution, prepared in PBS/BSA. Use PAG of defined size (e.g. 5 ± 1 nm in diameter), differing by at least 4 nm from that used in step 1. Note: the PAG solutions have to be centrifuged for 3 min at ~ 20000 *g* and room temperature to remove larger particle clusters immediately prior to use
9. Continue from step 9 of *Table 4.33*

Table 4.35. Pre-embedding procedure

1. Incubate cells or cellular components with the specific antibody (in excess) for 2 h at room temperature
2. Wash the sample twice by centrifugation and resuspension in an appropriate buffer (e.g. 50 mM PBS, pH 7.0–7.5 for cells or components where no osmotic stabilization is necessary; avoid conditions that weaken antibody binding such as low pH or high salt concentration) to remove unbound antibody
3. Incubate with the electron-dense marker system (e.g. Protein A coupled gold or ferritin coupled with a secondary antibody) for 2 h at room temperature. The optimum concentration is generally 10-fold higher than that used in post-embedding experiments (see above)
4. Proceed from step 1, *Table 4.16*, with the standard embedding procedure or perform deep-temperature embedding in Lowicryl resin, if additional post-embedding localization is desired (*Table 4.19*, step 2). After glutaraldehyde fixation, antigen–antibody binding is irreversible

Table 4.36. Whole-mount procedure

1. Allow the cell components to adsorb directly on to an electron-microscopic support film: place carbon-coated Formvar grids on the surface of a drop of the sample for 30–45 sec
2. Blot off the liquid on the grid surface completely or partially, depending on the particle concentration in the sample or the desired particle density. Isolated particles (e.g. membrane vesicles) not attached to each other and equally scattered over the grid surface should be preferred in order to achieve significant labelling signals
3. Before the liquid is completely removed, place the grid on to a drop of the IgG solution
4. Proceed as described for ultra-thin sections in *Table 4.33*, or, for double labelling, *Table 4.34*, performing washing steps after the antibody and protein–gold incubation very gently by placing the grid successively on the surfaces of two drops of PBS solution containing 0.01% (v/v) Tween and a third drop of PBS without Tween (no "jet wash")

Generally, all specific and non-specific binding reactions obtained in the whole-mount procedure are higher than those obtained during immunocytochemical localization with ultra-thin sections. The presence of casein hydrolysate (up to 2% w/v, final concentration) in the antibody and colloidal gold solution prior to the incubation steps serves as an appropriate quenching agent. Generally, higher dilutions of the antibody and especially of colloidal gold are recommended for the whole-mount procedure. Staining should be performed with sodium phosphotungstate (3%, w/v, phosphotungstic acid adjusted to pH 7.0 with NaOH solution).

Table 4.37. Immunolocalization procedure for cryosections

1. Warm the gelatin up to room temperature and then leave the floating grids for a further 10 min incubation
2. If the samples were subject to chemical fixation prior to rapid freezing, incubate on three drops of 20 mM glycine in PBS for 10 min each. Glycine will react with excess of aldehyde groups that may otherwise interfere with the protein solutions used in the subsequent incubation steps
3. Carefully remove the surplus liquid (do not blot dry) and transfer the grids on to drops of dilutions of the specific antibody (prepared in PBS containing 1% w/v BSA). Incubation time: 1–2 h
4. Incubate four times for 1 min on drops of PBS/BSA
5. Incubate on drops of the appropriate colloidal gold marker dilution (e.g. PAG prepared in PBS/BSA) for 30 min
6. Incubate four times for 5 min on drops of PBS. For most applications, the binding of antibodies and colloidal gold marker are stable enough to withstand the subsequent procedures. If not, incubate for 10 min on drops containing 1% glutardialdehyde in PBS (without BSA!). This step will cause covalent bonds to form between antigen, primary antibody and the colloidal gold-coupled protein
7. Incubate three times for 1 min on drops of distilled water and blot dry after the final incubation step
8. Stain the sections as described in *Table 4.27*

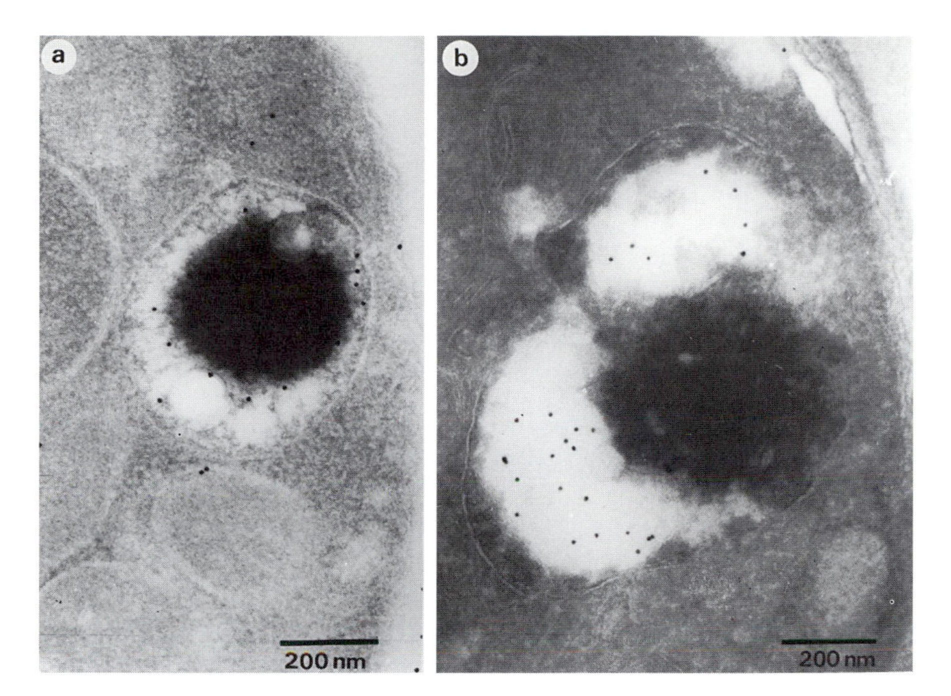

Figure 4.29. Appearance of a fungal vacuole after deep-temperature embedding in Lowicyl resin (see Section 4.6.2) and cryopreparation according to Tokuyasu (1986; see Section 4.6.5), after immunogold localization of vacuolar constituents. (a) Resin section, stained with 4% (w/v) uranyl acetate. (b) Cryosection stained with 0.3% (w/v) uranyl acetate/methylcellulose. Note the clear, sharp appearance of membranes in (b) as compared to in (a).

the grid successively on the surfaces of drops of the buffer solution. Double labelling and labelling with colloidal gold coupled with lectins or other markers may also be performed. Blocking of the sections to avoid non-specific reactions must be enhanced. A complete outline is given in *Table 4.37*. For the cryosectioning method, see *Table 4.26*.

4.7.3 Control experiments in immunocytochemistry

In order to exclude false negative or positive results in immunolocalization experiments, several controls are necessary. These controls check the specificity of the antibody–antigen binding as well as the binding of the marker system to the primary antibody:

- Incubate the specimen exclusively with the marker system. Any resulting binding is non-specific binding of the marker system to the sample. If this occurs, use a different marker system or try to reduce non-specific binding by the addition of quenching agents (e.g. casein hydrolysate) to the solution.
- Incubate the primary antibody with the respective purified antigen. The specimen should not subsequently be labelled by this mixture.

Labelling of the specimen indicates the presence of antibodies directed against other antigens or that bind non-specifically. Remove these antibodies by affinity chromatograpy with the antigen as ligand (specific antibody binds and has to be eluted in a second step) or use monoclonal antibodies.

- Incubate the specimen with the primary antibody, followed by an incubation with the secondary antibody or Protein A alone, followed by an incubation with the antibody or Protein A coupled with the electron-dense marker, respectively. If the marker system interacts specifically with the primary antibody, no labelling should occur. Avoid non-specific binding by using a different marker system or try to reduce non-specific binding by the addition of quenching agents (e.g. casein hydrolysate) to the solution.
- Incubate the sections with pre-immune serum (or any serum that does not contain the primary antibody), followed by the marker system. This reveals non-specific binding of antibodies to the specimen. Try to reduce non-specific binding by the addition of quenching agents (e.g. casein hydrolysate) to the solution; use monoclonal antibodies.
- Use a specimen that does not contain the antigen of interest. It is most appropriate if the same organism is used for the specific labelling and for the control experiment, but for the control to be grown under conditions where the cell component to be localized is not expressed. In cases where cell-wall labelling in Gram-negative bacteria or fungi is carried out with polyclonal antisera, it is strongly recommended that the specificity of the labelling is checked with other Gram-negative bacteria or fungi that do not contain the epitopes of interest.

4.7.4 Labelling of epitopes on isolated protein molecules

Localization of epitopes on the surface of (large) protein molecules may be performed with the aid of specific monoclonal or polyclonal antibodies directed against the epitope(s) in question (*Figure 4.26*). The protein sample is mixed with a monoclonal or polyclonal antibody at a concentration to give binding site equivalence. The sample is then incubated (a first experiment could entail an incubation time of 1 h at room temperature) and diluted by a factor of 10 to 100 with an appropriate buffer solution (e.g. 20 mM Tris–HCl, pH 7.0–7.5) and specimens are prepared by negative staining. Alternatively, after incubation, the antibody–protein complexes may be separated from unbound protein and antibody molecules by gel permeation chromatography prior to sample preparation. The antibody fraction may be replaced by Fab fragments or gold-labelled Fab fragments of antibodies (see Hermann *et al.*, 1991, and references cited therein for some applications).

4.8 Imaging and image evaluation with the transmission electron microscope

For imaging samples prepared as described in this book, a standard transmission electron microscope is used. Specimens may be visualized using different imaging modes (*Table 4.38*), most of which are available with the majority of transmission electron microscopes.

Transmission electron microscopes are available with different types of special accessories. A survey is given in *Table 4.39*.

Energy-filtering transmission electron microscopes (EFTEMs) facilitate an enormously increased range of applications. Since their range of applications is not restricted to using specially prepared samples (as e.g. cryo-stages are), a short summary of the basic features is given here. A built-in electron spectrometer (either "in-column", i.e. placed between the projective lens systems, or "post-column") divides the electron beam after its passage through the specimen into a spectrum of electrons with different energy losses. The energy losses derive from the interactions of the electrons with the sample and range from zero (no interaction) to energy losses of several hundreds of electron volts (and higher, but without relevance for EFTEMs), representing element specific energy losses (Bauer, 1988). By filtering these inelastically scattered electrons, EFTEMs allow unstained sections to be inspected, studies with very thick (around 1 μm) sections, compensation for undesired image contrast, detection of elements for contrast enhancement of specimens or markers (e.g. boron, see above) and elemental mapping of samples.

4.8.1 Stereo images

Production of stereo images requires a goniometer stage, which is useful for gaining insight into the third dimension of samples. In conventional TEMs the extension of the third dimension is restricted to the relatively low specimen thickness allowed, which should not exceed 150–200 nm, if a small objective aperture is used. Transmission of thicker sections is achieved by selection of higher accelerating voltages (100–500 eV) or filtering of inelastically scattered electrons in EFTEMs. Besides several other features (see below), EFTEMs permit examination of 1 μm thick specimens without considerable loss in image quality, that is, dimensions that have previously been restricted to high-voltage electron microscopes (Martin, 1996; Körtje *et al.*, 1996). Three-dimensional reconstruction by electron-microscopic tomography is facilitated by EFTEMs because of the increase in image contrast (Olins *et al.*, 1993, 1994).

Stereo images for simple illustration of three-dimensional structure do not need extensive adjustment procedures (see *Table 4.40*). Routinely, medium and high magnifications require a tilt angle of 12°, that is, the

Table 4.38. Imaging modes in standard TEM

Imaging/ operation modes[a]	Description	Comments
Bright field	Standard imaging mode: scattered and unscattered electrons participate in the formation of the image; the total electron dose specimens are subjected to is typically in the region of 1000 electrons $Å^{-2}$	Routinely used for stained specimens (ultra-thin sections, negative stained, metal shadowed)
Low-dose bright field achieved in connection with:	Total electron dose reduced to ≤ 10 electrons $Å^{-2}$	Used to reduce beam-induced damage of the specimen (e.g. with negatively stained protein) in the electron beam
minimal dose focusing	Focusing is carried out on an area different to that of interest (but in close proximity to it) by deflection of the electron beam	
Dark field	Only scattered electrons participate in the formation of the image; electron doses required are as much as ~30 times higher (!) than in bright-field mode (cf. Misell, 1978)	Visualization of e.g. nucleic acids in unstained form
Diffraction mode	Causes a magnified (magnification determined by the camera length) electron-diffraction pattern, which is present at the back focal plane of the objective lens, to be projected on to the screen	Used to record electron-diffraction patterns of crystalline specimens; contains high-resolution amplitude information which replaces the usually lower-resolution amplitude information present in the image and is combined with the image phase information (used in materials science and biological high-resolution TEM; cf. Misell and Brown, 1987)
Tilt series	Using an eucentric goniometer, the specimen is tilted relative to the incident electron beam	Tilted projections of single molecules (Osmani *et al.*, 1984), crystals and (thick) sections provide three-dimensional structural information (cf. Amos *et al.*, 1982; Engelhardt, 1988; Heppelmann *et al.*, 1989)

[a]A comprehensive description of electron microscope operation is given in Agar *et al.* (1987).

images of the stereo pair should be taken at –6°, and at +6° relative to the untilted position. This technique is very useful for evaluating marker (ferritin or colloidal gold) distribution after pre-embedding and

Table 4.39. Special equipment for electron microscopes

Special feature/ accessory	Description	Useful for
Cryo-stage/ cryo-holder	Stage/holder which can be operated with the specimen at liquid nitrogen (or possibly at liquid helium) temperatures	Observation of frozen-hydrated specimens
Electron energy loss spectroscopy (EELS)	Built-in energy-selecting electromagnetic prism sorts electrons according to their specific energy loss upon interaction with the specimen; electrons are analysed in a spectrometer	Element spectrum of small areas of the specimen
ESI	Equipment as described for EELS, but the electrons are used for imaging	Visualization of element distribution in ultra-thin sections and negatively stained specimens
X-ray element analysis	Analysis of element-specific X-rays emitted from the specimen on irradiation with electrons (requires specially equipped scanning electron microscopes or scanning transmission electron microscopes)	Visualization of the element distribution within specimens

Table 4.40. Production of stereo images

1. Select the desired area
2. If a eucentric goniometer stage is not used, rotate the grid until the desired area coincides with the position of the tilt axis
3. Tilt the specimen, reposition the desired area and readjust focus position. If an exact correction of magnification is necessary, report change in objective lens current. For a pair of stereo images, use tilt angles of –6° and +6°. If parallactic shift exceeds 0.5 cm in the final image, lower tilt angles must be used (see footnote)

The actual magnification obtained depends on the current of the objective lens. When the area is positioned in the tilt axis, refocusing is not necessary or, at least, kept to a minimum. If refocusing is necessary, the difference in the objective lens currents has to be determined and used for recalculation of the actual magnification. (The formula depends on the instrument type and is included in the respective instruction manuals.)

Correction of different magnifications in a stereo pair may also be achieved during processing of photographic prints. Since parallactic shift parallel to the tilt axis is zero, the distance between two points on a line parallel to the tilt axis does not change. Therefore, magnifications of reproductions should be adjusted in such a way that these points are the same distance apart in a stereo pair. In order to achieve the stereo effect, parallactic shifts in the final image pair should not exceed 0.5 cm.

post-embedding labelling (*Figure 4.30*) of specimens. Localization of marker positions relative to cellular components in thick sections requires a three-dimensional image. Stereological measurements require more precise adjustments (see Helmcke, 1980).

Three-dimensional image reconstruction of complex tissue from serial sections by application of conventional TEM is time consuming, since sections of 80 nm in thickness have to be used for high-resolution

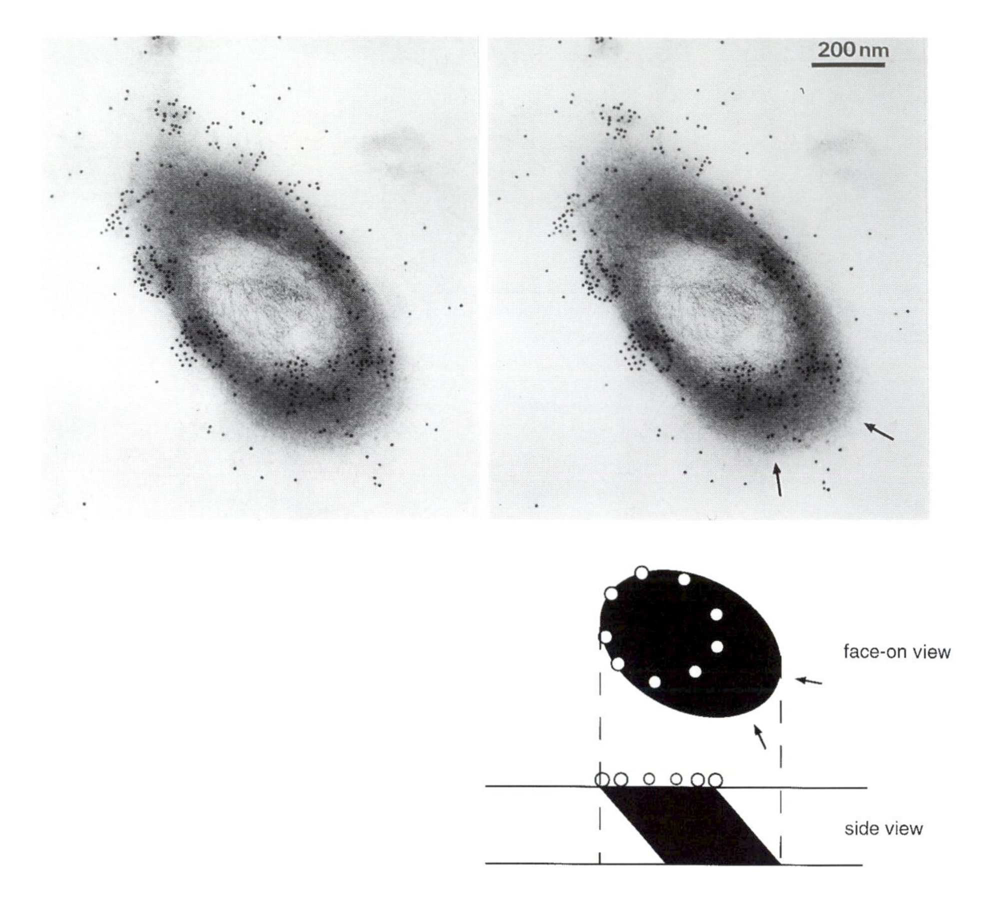

Figure 4.30. Stereoview of a post-embedded cell (section thickness: 500 nm). Immunogold labelling of the bacterial capsule. Not all gold markers seem to be located at the cell periphery, but the stereoview reveals that the cell section is oriented in the resin section as indicated by the schematic drawing, leading to the observed distribution of markers. (Original micrograph: S. Vetterleund.)

images. Thick sections allow fast reconstruction of three-dimensional views with only minimal loss in resolution when ESI is applied (Heppelmann *et al.*, 1989).

In principle, tilted images obtained from serial sections may be analysed and recalculated to a three-dimensional model of subcellular structures, when appropriate image processing is applied (Olins *et al.*, 1994).

5 Methods for scanning electron microscopy (SEM)

Scanning electron microscopy is very useful for investigating the general shape and surface morphology of bacterial cells and their interaction with each other as well as their environment (cf. Hollenberg and Erickson, 1973; Reimer, 1978). In essence, a focused electron beam is scanned across the sample, causing the release of backscattered and secondary electrons from the surface. These electrons are collected and a magnified three-dimensional image is formed which is displayed on a cathode-ray monitor. Owing to their great depth of focus, SEM images can usually be interpreted in the same way as any other three-dimensionally perceived objects.

During recent years, advances in instrumentation (low temperature SEM, field emission guns; e.g. Hermann and Müller, 1991; Walther *et al.*, 1995) and specimen preparation (see below) have significantly improved the fidelity and resolution of SEM studies, enabling even molecular assemblies (e.g. viruses) to be visualized.

5.1 Standard specimen preparation

For visualization in SEM, it is necessary that specimens emit a sufficient number of secondary electrons and develop as few surface charges as possible in the electron beam. Only a few biological samples, such as insect exoskeletons, bone or diatom frustules, fulfil these prerequisites. In most cases, special preparation procedures are necessary. For further information on the preparation of bacteriological specimens, the reader is referred to Watson *et al.* (1980) and the fine selection of SEM-specific references quoted by Holt and Beveridge (1982). More recent developments, including low-temperature SEM, are described by Read and Jeffree (1991).

5.1.1 Preparation of a specimen for SEM

Fixation. Fixation can be achieved using either conventional or cryo methodology. Conventional (chemical) fixation of the sample is carried out in an isotonic medium (the media are described in more detail for fixation of samples to be embedded in resin; see section 4.6.1) containing 1–4% (v/v) glutardialdehyde. Since osmium tetroxide considerably increases the conductivity of the specimen (and thereby gives rise to a higher emission of electrons), a second fixation step is usually recommended: the sample is subjected to alternate incubation procedures in osmium tetroxide (1–2%, w/v) and tannic acid (1–7%, w/v) or thiocarbohydrazide (saturated solution). During this procedure, a film of osmium is deposited on the surface of the specimen. Therefore, the objects often do not need further metal coating. A different strategy involving the fixation of single bacterial colonies in osmium tetroxide vapours has been devised by Springer and Roth (1972) and consists of the steps listed in *Table 5.1*.

The block is now ready to be processed further. Following this particular protocol, the agar blocks are glued on to stubs for drying in air or in a desiccator, which is different from the ways described below, but worth considering when viewing colonies of microbial cells. However, the undisputedly most gentle and efficient way to an artefact-free fixation of samples is by plunging them into liquid cryogen (physical fixation; see above) after the cells have been adsorbed on suitable specimen supports, such as, for example, carbon-coated copper grids or carbon films of 0.1–0.2 mm thickness. The latter films are self-supporting and are purchased as larger sheets (e.g. from Goodfellow Cambridge Ltd., UK) from which smaller areas can be cut off to suit the sample. When adsorbing filamentous or network-forming microorganisms on to a support, one ought to be aware that any shearing forces resting on the sample may introduce artefacts by disrupting the original organization of the colony. In cases where this organization matters (e.g. in cases where an organization into mats, fibres, etc., is macroscopically evident), one should, therefore, avoid centrifugation steps and transfer them directly from the medium on to the support (see *Figure 5.1*).

Table 5.1. Osmium staining for SEM

1. Cut out a small block (~2 mm height) of agar supporting a single colony
2. Place the agar block in a Petri dish at the periphery and incline the dish so that the cube will be located at the highest point
3. Place a few drops of OsO_4 in the dish, cover with a lid, seal tightly and incubate overnight
4. Using a pipette, remove the OsO_4 and carefully rinse the lower area with water
5. Level the dish out on a bench and surround the block with water, always making sure that no water ever contacts the upper surface of the block. Incubate for 15 min and repeat twice. This step is carried out to decrease the salt concentration in the agar block
6. Excess of liquid is removed by transferring the block on to filter paper

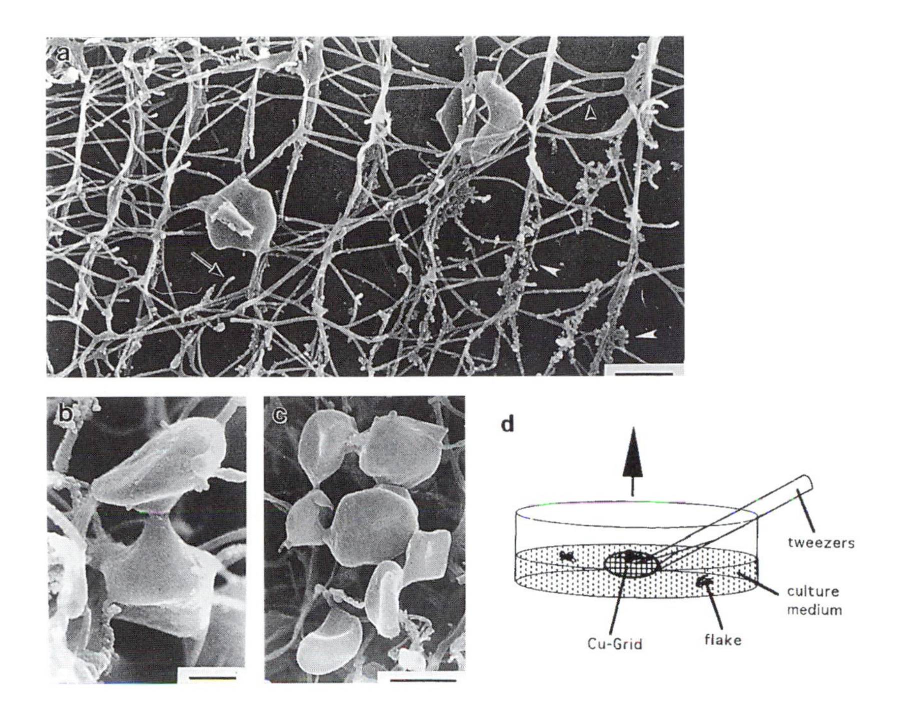

Figure 5.1. SEM images of *Pyrodictium abyssi.* (a) *Pyrodictium* cells grow in flakes consisting of a cobweb-like extracellular matrix built up by tubules in which the cells are entrapped (1–10 nm in diameter; the arrow marks a "dead end" of a tubule, the arrowhead a knot). Bar represents 1 μm. (b) Cells, most likely in a late phase of division. Bar represents 0.5 μm. (c) Small aggregate. Bar represents 1 μm. (d) Collecting the flakes out of the culture medium on to a copper grid. From Rieger *et al.* (1995). With kind permission of Academic Press Inc. and the authors R. Rachel and R. Hermann.

Dehydration and drying. The chemically fixed specimens are then subjected to dehydration in a graded series of acetone or ethanol, as described for the dehydration of samples for resin embedding and ultramicrotomy (see above). Physically fixed specimens are sensibly subjected to freeze substitution (see above). Freeze substitution against acetone can be carried out using 2% osmium tetroxide solution as fixing additive for 3 × 8 h at –90°, –60° and –30°C, respectively (Rieger *et al.*, 1995). After the water has been removed from the medium, the organic solvent also needs to be removed. In order to avoid drying artefacts, this is done by critical-point drying or freeze drying. In most cases, critical-point drying is the preferred method since it is cheap and fast. In brief, the specimen is transferred from the dehydration fluid (ethanol) into the drying medium (gaseous CO_2) via a transitional fluid (liquid CO_2). Exchange of the transitional fluid for the drying medium is done under high pressure (critical pressure) and at the critical temperature of the transitional fluid (31°C for CO_2, but 374.2°C for water, which explains why it is not suitable). Above the critical point, the densities of the drying medium in its liquid and gas phases are identical. If the temperature of the system remains above the critical point, the gas phase can be vented off without

Table 5.2. Principal steps involved in critical-point drying

1. Precool the apparatus to 5–10°C below the ambient temperature by flushing with liquid CO_2 (modern pieces of apparatus are likely to be equipped with cooling coils, usually obviating the need for this flushing step)
2. Load samples immersed in ethanol (or acetone) and seal the chamber
3. Open the CO_2 inlet valve and allow the chamber to fill with CO_2
4. Slightly open the exhaust valve to expel any ethanol
5. Close the exhaust valve and fill the chamber with CO_2 again
6. Close the inlet valve and allow the specimen to remain in liquid CO_2 for ~ 30 min, to allow time for the CO_2 to replace any dehydration fluid
7. Flush again with CO_2, until there is no ethanol left
8. Close both valves
9. Turn on the heater and allow the temperature to rise above 31°C and the pressure to build up to 1100–1200 p.s.i.
10. Once the critical temperature and pressure are exceeded, slowly vent off the CO_2, while keeping the temperature elevated
11. The specimens are now ready to be mounted on stubs and sputter coated

subsequent recondensation. Because CO_2 is removed after its transition from the liquid to the gas phase at the critical point, the specimen is dried without structural damage. After removal of the gas and decompression of the chamber, the dried specimen is transferred to a desiccator (see *Table 5.2* for details).

Sputtering. Conductivity of (biological) specimens has to be increased prior to visualization in the electron microscope. The fixation procedure may provide sufficient conductivity (see above), but normally, sputtering (Au, W or Pt, the latter two have a smaller particle size) or vacuum evaporation of Au or Au–Pd is necessary.

Sputtering of the metal is done in a vacuum chamber equipped with a gold-plated cathode and a glow-discharge unit. Glow discharging results in the release of gold atoms which subsequently cover the sample (anode).

In addition to this standard procedure, special preparation methods such as fracturing of specimens, elemental analysis and immunocytochemical localization can be used in conjunction with SEM (see *Table 5.3*).

Table 5.3. Special methodical approaches utilizing SEM

Approach	Description	Useful for
Fracturing/cleaving of objects	Sectioned embedded material, freeze-fractured specimens or mechanically disrupted material is prepared for visualization in the SEM	Visualization of internal structures of e.g. wood, leaves or tissues (see, e.g. Hotta *et al.*, 1990)
Elemental analysis	Measuring of cathodoluminescence or X-rays caused by the interaction of the electron beam with the specimen	Recording spectra or images depicting the elemental distribution of the sample
Immunocytochemical localization	Localization of antigens with the aid of colloidal gold, subsequent preparation for SEM	Localization of antigens exposed to the surface of the specimen (Hodges for *et al.*, 1987)

6 Image evaluation in TEM and SEM

The qualitative and quantitative assessment of electron micrographs provides information not readily obtainable from a mere description of the object. Prior to this assessment, one should make sure that only optically sound electron micrographs depicting specimens free of obvious artefacts should be subjected to further analysis.

Processing of EM images, either as video signals or electron micrographs, in order to enhance or quantify signals is becoming increasingly important for interpretation of the results. Whereas in the past a wide variety of specialized (and expensive) tools (e.g. laser diffractometers, densitometers and stereocomparators) were helpful for image evaluation, nowadays most of these tasks are performed, often with greater accuracy, on a computer workstation or even on a small personal computer (PC). The methods presented here give only some application examples for evaluation methods that do not require specialized computer equipment or software and are restricted to processing methods that are useful in day-to-day EM.

Numerous simple statistical analyses such as counting particles or measurement of particle dimensions (e.g. surface dimensions of cells or subcellular components) can easily be carried out using only a ruler (or magnifying glass with a graticule) and a calculator, and quite often provide important additional information. Computer-assisted evaluation of data (via, e.g. a scanner and graphics tablet in connection with a PC) speeds up the analysis and improves the statistical analysis owing to an easier acquisition/analysis of larger data sets. As a first step in this direction if a PC is available, one might wish to access public domain software available for image analysis and processing (Lennard, 1990).

More sophisticated software is needed for the three-dimensional reconstruction of images of cells from ultra-thin sections, or for averaging and three-dimensional reconstruction of projections of biological macromolecules. *Table 6.1* provides a brief survey of some digital image analysis and processing methods. An excellent textbook covering the practical and mathematical background of image analysis from first principles in an easily understandable manner is the one by Misell (1978).

Table 6.1. Methods for the evaluation of electron microscopical data

Problem	Method	Comment/references
Evaluation of length, surface and volume	• Morphometric assessment with the aid of transparent test lattices (linear, point, grid system) • PC-supported assessment (also requires a graphics tablet, scanner or video camera, etc.)	Cheap methods for the statistical evaluation of data on cell dimensions, etc. (Loud and Anversa, 1984)
Three-dimensional information about cells and subcellular components	Computer-assisted calculation of • the three-dimensional shape from serial thin sections, • three-dimensional representations of thick sections and SEM images	(Baba *et al.*, 1989; Gremillet *et al.*, 1994)
Correlation averaging and three-dimensional reconstruction of single particles and two-dimensional (2-D) protein crystals (e.g. Henderson *et al.*, 1990; for introductory overviews on transmembrane proteins, see Holzenburg *et al.*, 1992 and Holzenburg, 1997)	Analysis and processing of digitized electron micrographs of single molecules/ two-dimensional crystals and tilt series with the aid of image-processing software systems; a brief overview of software packages, as well as names and addresses of suppliers, is provided by Hegerl (1992); further packages are AIX (Lanzavecchia *et al.*, 1993), CRISP (DOS™ and WINDOWS™ compatible; Hovmöller, 1992) and SPECTRA (Schmid *et al.*, 1993)	Special hardware/software and experience necessary (for a bibliographic survey, see Chiu *et al.*, 1993; N.B. *The Journal of Structural Biology* has published a special issue (vol. 116, no. 1) dedicated to "Advances in computational image-processing for microscopy. A comprehensive treatise") Protocols for the production of 2-D crystals have been devised for soluble proteins (Holzenburg, 1988; Harris and Holzenburg, 1995), transmembrane proteins (Holzenburg, 1995 and references therein), and membrane-associated proteins (Brisson *et al.*, 1994). 2-D and 3-D crystals can also be analysed by combining X-ray with EM data (Holzenburg *et al.*, 1987; Glykos *et al.*, 1998).

The following protocols refer to image processing and analysis methods that are very useful in the interpretation and statistical evaluation of electron micrographs but do not need specialized (and expensive) image-analysis software. The equipment required comprises

- Standard darkroom equipment.
- Personal computer (IBM compatible, Apple Macintosh™ or equivalent, at least 16 Mbytes of random access memory).

Table 6.2. Particle counting and measurement of surface areas (commands are in italics)

1. Extract the desired area as follows: apply *threshold* to separate areas of defined grey levels. This method is suitable for extracting the area of a whole embedded cell or a vacuole or storage granule inside a cell. Determine the surface area
2. Apply *threshold* to extract the dark dots representing colloidal gold particles. Since the particles have a defined size, smaller or larger dark areas may be removed by exclusion of areas below a minimal and above a maximal size
3. *Analyse* particles by determination of the particle number. Together with the surface area from step 1, the density of gold particles may be determined

- Equipment for image acquisition (e.g. scanner, video camera and appropriate software).
- Image processing software.

6.1 Quantification of results (surface area, particle counting)

Some simple analysis methods permit the quantification of immunolocalization results (*Table 6.2*; *Figure 6.1*). The protocol refers to commands that are common to several algorithms, mainly within the freely accessible "Image" software (Apple Macintosh™ required; software available via the Internet: ftp://zippy.nimh.nih.gov/pub/nih-image/).

6.2 Densitograms

Densitometry is very helpful when membrane dimensions have to be illustrated and quantified. Options for data evaluation by densitometry are regularly included in protein and DNA gel analysis software and are, thus, present in most biomedical laboratories. Determination of membrane thickness from densitograms may be achieved by measuring the straight-line distance between the two inflection points of a peak (or double peak) that represents the membrane. The inflection points correspond to the maxima and minima in the first derivative of the original plot (see *Figure 6.2*; Mayer and Hoppert, 1997).

6.3 Markham rotation

Markham rotation (Markham *et al.*, 1963) is a simple method to confirm or determine rotational symmetries of biological structures. The original

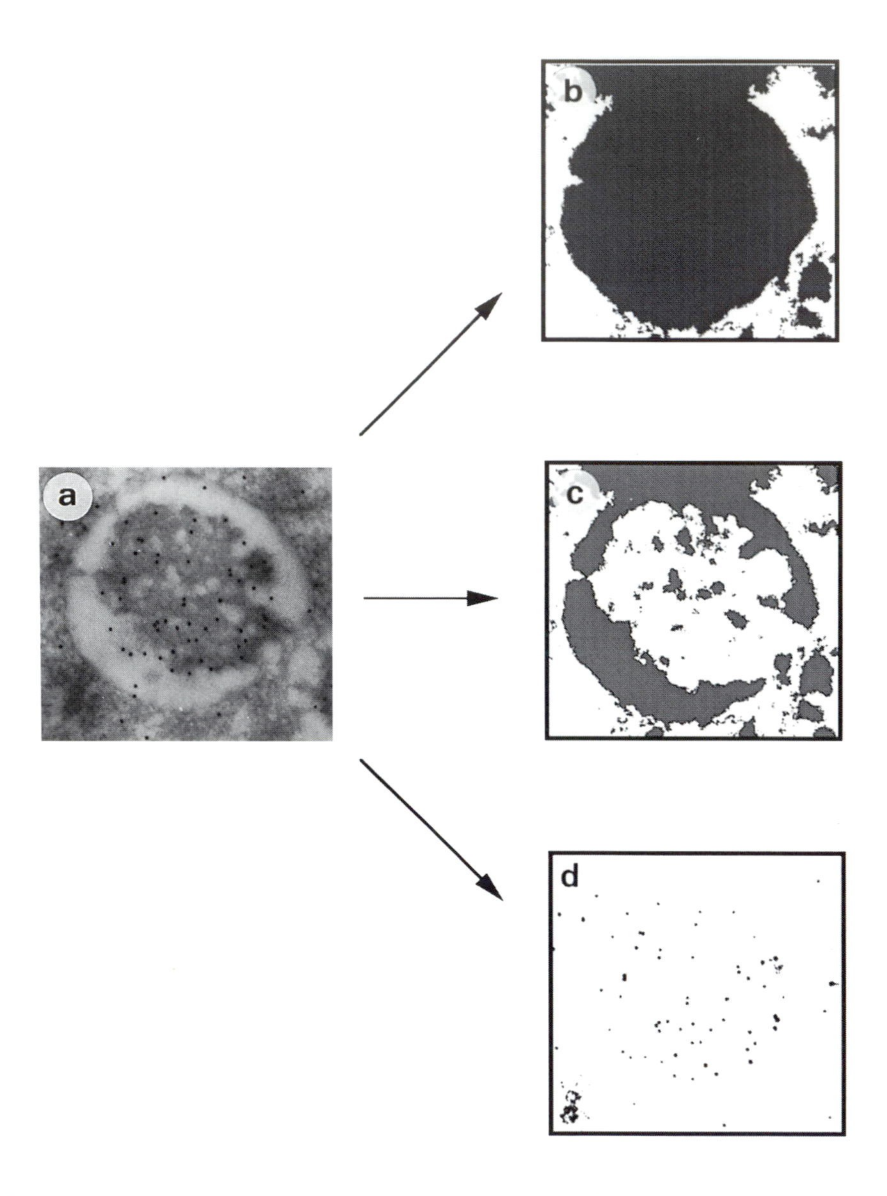

Figure 6.1. Quantitation of immunolocalization data. (a) Original electron micrograph; fungal vacuole, containing a heterogeneously stained inclusion body, densely labelled with immunogold markers. Problem: determination of the surface area of the vacuolar section, and determination of marker density (markers/defined area) in the dark stained inclusion. The area of the inclusion body itself cannot be determined, because discrimination between the inclusion body and the cytoplasm is not possible by thresholding, i.e. extraction of grey levels. (b) Determination of surface area within a closed boundary line represented by the vacuolar membrane (dark: extracted area, i. e. the area of the vacuole). (c) Thresholding and determination of the light (i.e. unstained) area of the vacuole, which is free of markers. Area determined in (b) *minus* area determined in (c) represents the area of the inclusion body. (d) Thresholding and counting closed areas within a defined size range above a defined grey level (i.e. gold markers of 10 nm).

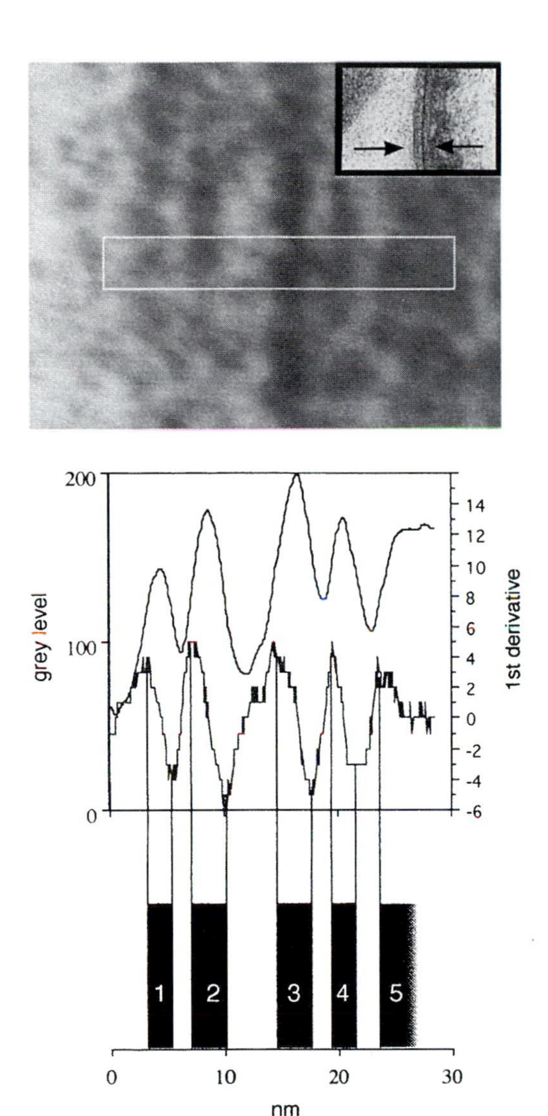

Figure 6.2. Densitogram of the cell periphery from *Ralstonia eutropha* (as depicted in *Figure 4.22c*). The original electron micrograph (see inset) was digitized. Determination of grey levels has been made across the boxed area (upper curve). The lower curve depicts the first derivative of the grey level curve, allowing the determination of the inflection points of the grey level curve. The distances between the turning points represent the dimensions of the membrane leaflets as depicted in the schematic drawing. 1: Outer membrane (stained outer leaflet); 2: outer membrane (stained inner leaflet); 3: peptidoglycane layer; 4: cytoplasmic membrane (stained outer leaflet); 5: cytoplasmic membrane (stained inner leaflet). See also Mayer and Hoppert (1997).

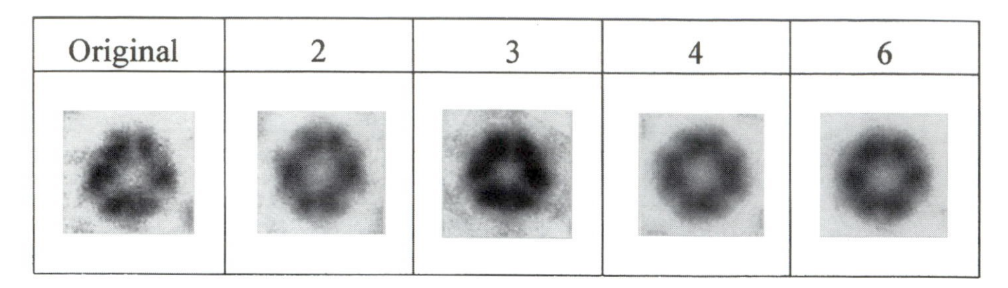

Figure 6.3. Markham rotation of a large enzyme molecule (same specimen as in *Figure 4.1,* projection form (a)). The rotation of the original image by 120° and 240° and the subsequent superimposition (three subdivisions of a circle) enhance the contrast. The other images appear blurred. Original micrographs: I. J. Braks.

images are rotated by increments of a circle. If n meets a periodicity of the structure, the resulting superimposed image will reflect this periodicity by revealing enhanced contrast. Blurred images indicate the absence of increments of rotational symmetry.

Markham rotation may be performed in the darkroom by multiple (n-fold) exposure of a photographic print that is rotated by n increments with the point of n-fold rotational symmetry as rotational axis. When appropriate software is present (e.g. the readily available Adobe Photoshop™ software for image processing), rotation of images by defined angles and subsequent superposition of the image stack may be performed within a few minutes (see *Figure 6.3*).

Appendix A

References and further reading

References

Abermann, R. and Bachmann, L. (1969) Elektronenmikroskopische Beschattung mit hoher Auflösung. *Naturwissenschaften* **56**: 324.

Abram, D., Vatter, A.E. and Koffler, H. (1966) Attachment and structural features of flagella of certain bacilli. *J. Bacteriol.* **91**: 2045–2068.

Acker, G. (1988) Immunoelectron microscopy of surface antigens (polysaccharides) of gram-negative bacteria using pre- and post-embedding techniques. In: *Methods in Microbiology*, Vol. 20: *Electron Microscopy in Microbiology* (ed. F. Mayer). Academic Press, London, pp. 147–174.

Adrian, M., Dubochet, J., Lepault, J. and McDowall, A.W. (1984) Cryo-electron microscopy of viruses. *Nature* **308**: 32–36.

Agar, A.W., Alderson, R.H. and Chescoe, D. (1987) Principles and practice of electron microscope operation. In: *Practical Methods in Electron Microscopy*, Vol. 2 (ed. A.M. Glauert). Elsevier, Amsterdam.

Aldrich, H.C., McDowell, L., Barbosa, M.F., Yomano, L.P., Scopes, R.K. and Ingram, L.O. (1992) Immunocytochemical localization of glycolytic and fermentative enzymes in *Zymomonas mobilis*. *J. Bacteriol.* **174**: 4504–4508.

Allen, E.D. and Wheatherbee, L. (1979) Ultrastructure of red cells frozen with hydroxyethyl starch. *J. Micros.* **117**: 381–394.

Almeida, J.D. (1980) Practical aspects of diagnostic electron microscopy. *Yale J. Biol. Med.* **53**: 5–18.

Amos, L.A., Henderson, R. and Unwin, P.N.T. (1982) 3-dimensional structure determination by electron microscopy of two-dimensional crystals. *Prog. Biophys. Mol. Biol.* **39**: 183–231.

Aragno, M., Wather-Mauruschat, A., Mayer, F. and Schlegel, H.G. (1977) Micromorphology of gram-negative bacteria. I. Cell morphology and flagellation. *Arch. Microbiol.* **114**: 93–100.

Baba, N., Baba, M., Imamura, M., Koga, M., Ohsumi, Y., Osumi, M. and Kanaya, K. (1989) Serial section reconstruction using a computer graphics system: applications to intracellular structures in yeast cells and to the peridontal structure of dog's teeth. *J. Elect. Micros. Tech.* **11**: 16.

Bachmann, L. and Schmitt, W.W. (1971) Improved cryofixation applicable to freeze-etching. *Proc. Natl Acad. Sci. USA* **68**: 2149–2152.

Bachmann, L., Orr, W.H., Rhodin, T.N. and Siegel, B.M. (1960) Determination of surface structure using ultra-high vacuum replication. *J. Appl. Phys.* **31**: 1458–1462.

Bauer, R. (1988) Electron spectroscopic imaging: an advanced technique for imaging and analysis in transmission electron microscopy. In: *Methods in Microbiology*, Vol. 20 (ed. F. Mayer). Academic Press, London, pp. 113–146.

Beesly, J.E. (1989) Colloidal gold: A new perspective for cytochemical marking. *Microscopy Handbooks*, Vol. 17. Royal Microscopic Society and Oxford University Press, Oxford.

Bendayan, M. (1985) The enzyme–gold technique: a new cytochemical approach for the ultrastructural localization of macromolecules. In: *Techniques in Immunocytochemistry*, Vol. 3 (eds. G.R. Bullock and P. Petrusz). Academic Press, London, pp. 179–201.

Bendayan, M., Barth, R.F., Gingras, D., Londono, I., Robinson, P.T., Alam, F., Adams, D.M. and Mattiazzi, L. (1989) Electron spectroscopic imaging for high-resolution immunocytochemistry: use of boronated Protein A. *J. Histochem. Cytochem.* **37**: 573–580.

Birkenhäger, R., Hoppert, M., Deckers-Hebestreit, G., Mayer, F. and Altendorf, K.-H. (1995) The Fo complex of ATP synthase of *Escherichia coli*: Investigation by electron spectroscopic imaging and immunoelectron microscopy. *Eur. J. Biochem.* **230**: 58–67.

Blöchl, E., Rachel, R., Burggraf, S., Hafenbradl, D., Jannasch, H.W. and Stetter, K.O. (1997) *Pyrolobus fumarii*, gen. and sp. nov. represents a novel group of archaea, extending the upper temperature limit for life to 113°C. *Extremophiles* **1**: 14–21.

Bodenstein, C. (1991) *Untersuchungen zur Membranverankerung der membrangebundenen Hydrogenase aus Alcaligenes eutrophus H16 PHB-4.* Ph.D thesis, Univ. Göttingen.

Bradley, D.E. (1954) Evaporated carbon films for use in electron microscopy. *Br. J. Appl. Phys.* **5**: 65–66.

Bradley, D.E. (1958) Simultaneous evaporation of platinum and carbon for possible use in high-resolution shadow-casting for the electron microscope. *Nature* **181**: 875–877.

Braks, I.J., Hoppert, M., Roge, S. and Mayer, F. (1994) Structural aspects and immunolocalization of the F_{420}-reducing and non-F_{420}-reducing hydrogenases from *Methanobacterium thermoautotrophicum* Marburg. *J. Bacteriol.* **176**: 7677–7687.

Brenner, S. and Horne, R.W. (1959) A negative staining method for high resolution electron microscopy of viruses. *Biochim. Biophys. Acta* **34**: 103–110.

Brisson, A., Olofsson, A., Ringler, P., Schmutz, M. and Stoylova, S. (1994) Two-dimensional crystallization of proteins on planar lipid films and structure determination by electron crystallography. *Biol. Cell.* **80**: 221–228.

Carlemalm, E., Garavito, R.M. and Villiger, W. (1982) Resin development for electron microscopy and an analysis of embedding at low temperature. *J. Micros.* **126**: 123–143.

Chiu, W., Schmid, M.F. and Venkataram Prasad, B.V. (1993) Teaching electron diffraction and imaging of macromolecules. *Biophys. J.* **64**: 1610–1625.

Däkena, P., Rohde, M., Dimroth, P. and Mayer, F. (1988) Oxaloacetate decarboxylase from *Klebsiella pneumoniae*: size and shape of the enzyme, and localization of its prosthetic group by electron microscopic affinity labeling. *FEMS Microbiol. Lett.* **55**: 35–40.

Devereux, J., Haeberli, P. and Smithies, O. (1984) A comprehensive set of sequence analysis programs for the VAX. *Nucleic Acids Res.* **12**: 387–395.

Drummond, D.G. (1950) The practice of electron microscopy. *J. Roy. Micros. Soc.* **70**: 1–141.

Dubochet, J., Ducommun, M., Zollinger, M. and Kellenberger, E. (1971) A new preparation method for dark-field electron microscopy of biomacromolecules. *J. Ultrastruct. Res.* **35**: 147–167.

Dubochet, J., Lepault, J., Freeman, R., Berriman, J.A. and Homo, J.-C. (1982) Electron microscopy of frozen water and aqueous solutions. *J. Micros.* **128**: 219–237.

Dubochet, J., Adrian, M., Chang, J.-J., Homo, J.-C., Lepault, J., McDowall, A.W. and Schultz, P. (1988) Cryo-electron microscopy of vitrified specimens. *Quart. Rev. Biophys.* **21**: 129–228.

Egger, D., Troxler, M. and Bienz, K. (1994) Light and electron microscopic *in situ* hybridization: non-radioactive labeling and detection, double hybridization, and combined hybridization–immunocytochemistry. *J. Histochem. Cytochem.* **42**: 815–822.

Eiffert, H., Schlott, T., Hoppert, M., Lotter, H. and Thomssen, R. (1992) Identification of an endoflagellar associated protein in *Borrelia burgdorferi*. *J. Med. Microbiol.* **36**: 209–214.

Eismann, K., Mlejnek, K., Zipprich, D., Hoppert, M., Gerberding, H. and Mayer, F. (1995) Antigenic determinants of the membrane-bound hydrogenase in *Alcaligenes eutrophus* are exposed toward the periplasm. *J. Bacteriol.* **177**: 6309–6312.

Engelhardt, H. (1988) Correlation averaging and 3-D reconstruction of 2-D crystalline membranes and macromolecules. In: *Methods in Microbiology*, Vol. 20 (ed. F. Mayer). Academic Press, London, pp. 357–413.

Escaig, J. (1982) New instruments which facilitate rapid freezing at 83 K and 6 K. *J. Micros.* **126**: 221–230

Estis, L.F., Haschemeyer, R.H. and Wall, J.S. (1981) Uranyl sulphate: a negative stain for electron microscopy. *J. Micros.* **124**: 313–318.

Fabergé, A.C. and Oliver, R.M. (1974) Methylamine tungstate, a new negative stain. *J. Micros. (Paris)* **20**: 241–246.

Farrant, J.L. (1954) An electron microscopic study of ferritin. *Biochim. Biophys. Acta* **13**: 569–576.

Frens, G. (1973) Preparation of gold dispersions of varying particle size: controlled nucleation for the regulation of the particle size in monodisperse gold suspensions. *Nature Phys. Sci.* **241**: 20–22.

Freudenberg, W., Mayer, F. and Andreesen, J.R. (1989) Immunocytochemical localization of proteins P1, P2, P3 of glycine decarboxylase and of the selenoprotein P_A of glycine reductase, all involved in anaerobic glycine metabolism of *Eubacterium acidaminophilum*. *Arch. Microbiol.* **152**: 182–188.

Frösch, D. and Westphal, C. (1989) Melamine resins and their application in electron microscopy. *Electron Micros. Rev.* **2**: 231–255.

Geoghegan, W.D. and Ackerman, G.A. (1977) Adsorption of horseradish peroxidase, ovomucoid and anti-immunglobuline to colloidal gold for the indirect detection of concanavalin A, wheat germ agglutinin and goat antihuman immunglobuline G on cell surfaces at the electron microscopic level: a new method, theory and application. *J. Histochem. Cytochem.* **25**: 1187–1200.

Gerdes, H.H. and Kaether, C. (1996) Green fluorescent protein: applications in cell biology. *FEBS Lett.* **389**: 44–47.

Geymayer, W., Grasenick, F. and Hodl, Y. (1977) Stabilising ultrathin cryosections by freeze-drying. *J. Micros.* **112**: 39–46.

Glauert, A.M., Rogers, G.E. and Glauert, R.H. (1956) A new embedding medium for electron microscopy. *Nature* **178**: 803.

Glykos, N.M., Holzenburg, A. and Phillips, S.E.V. (1998) Low-resolution structural characterization of the arginine repressor/activator from *Bacillus subtilis*: a combined X-ray crystallographic and electron microscopical approach. *Acta Cryst.* **D 54**: 215–225.

Gregg, M. and Reznik-Schüller, H.M. (1984) An improved method for rapid electron microscopic autoradiography. *J. Micros.* **135**: 115–118.

Greipel, J., Maass G. and Mayer, F. (1987) Complexes of the single strand DNA binding protein from *E. coli* (EcoSSB) with poly(dT). An investigation of their structure and internal dynamics by means of electron microscopy and NMR. *Biophys. Chem.* **26**: 149–161.

Gremillet, P., Jourlin, M. and Pinoli, J.C. (1994) LIP-model-based three-dimensional reconstruction and visualization of HIV-infected entire cells. *J. Micros.* **174**: 31–38.

Griffiths, G., Quinn, P. and Warren, G. (1983a) Dissection of the Golgi complex. I. Monensin inhibits the transport of viral membrane proteins from medial to trans Golgi cisternae in baby hamster-kidney cells infected with Semliki forest virus. *J. Cell Biol.* **96**: 835–850.

Griffiths, G., Simons, K., Warren, G. and Tokuyasu, K.T. (1983b) Immunoelectron microscopy using thin, frozen sections: application to studies of the intracellular transport of Semliki forest virus spike glycoproteins. *Methods Enzymol.* **96**: 435–450.

Griffiths, G., Mcdowall, A.F., Back, R. and Dubochet, J. (1984) On the preparation of cryosections for immunochemistry. *J. Ultrastruct. Res.* **89**: 65–78.

Gross, H., Müller, T., Wildhaber, I. and Winkler, H.P. (1984) Recent progress in high-resolution shadowing. *Proc. 8th Eur. Congr. Elec. Micros.* 1293–1294.

Harris, J.R. and Holzenburg, A. (1995) Human erythrocyte catalase: 2-D crystal nucleation and production of multiple crystal forms. *J. Struct. Biol.* **115**: 102–112.

Harris, J.R. and Horne, R.W. (1994) Negative staining — a brief assessment of current technical benefits, limitations and future possibilities. *Micron* **25**: 5–13.

Haschemeyer, R.H. and Myers, R.J. (1972) Negative staining. In: *Principles and Techniques of Electron Microscopy*, Vol. 2 (ed. M. Hayat). Van Nostrand Reinhold, New York, pp. 101–147.

Hayat, M.A. (1973–1977) *Electron Microscopy of Enzymes: Principles and Methods*. Vol. 1 (1973); Vols 2 and 3 (1974); Vol. 4 (1976); Vol. 5 (1977). Van Nostrand Reinhold, New York.

Hayat, M.A. (1981) *Fixation for Electron Microscopy*. Academic Press, London.

Hegerl, R. (1992) A brief survey of software packages for image processing in biological electron microscopy. *Ultramicroscopy* **46**: 417–423.

Heinmets, F. (1949) Modification of silica replica technique for study of biological membranes and application of rotary condensation in electron microscopy. *J. Appl. Phys.* **20**: 384–389.

Helmcke, J.G. (1980) Die Erzeugung von Stereobildern und Stereogrammen. In: *Methodensammlung der Elektronenmikroskopie 4.4.2* (eds G. Schimmel and W. Vogell). Wissenschaftliche Verlagsgesellschaft, Stuttgart.

Henderson, R., Baldwin, J.M., Ceska, T.A., Zemlin, F., Beckmann, E. and Downing, K.H. (1990) Model for the structure of bacteriorhodopsin based on high-resolution electron cryo-microscopy. *J. Mol. Biol.* **213**: 899–929.

Heppelmann, B., Messlinger, K. and Schmidt, R.F. (1989) Serial sectioning, electron microscopy, and three-dimensional reconstruction of fine nerve fibres and other extended objects. *J. Micros.* **156**: 163–172.

Hermann, R. and Müller, M. (1991) Prerequisites of high-resolution scanning electron microscopy. *Scanning Micros.* **5**: 653–664.

Hermann, R., Schwarz, H. and Müller, M. (1991) High precision immunoscanning electron microscopy using Fab fragments coupled to ultra-small colloidal gold. *J. Struct. Biol.* **107**: 38–47.

Hippe-Sanwald, S. (1993) Impact of freeze substitution on biological electron microscopy. *Micros. Res. Tech.* **24**: 400–422.

Hodges, G.M., Southgate, J. and Toulson, E.C. (1987) Colloidal gold — a powerful tool in scanning electron microscope immunocytochemistry: an overview of bioapplications. *Scanning Micros.* **1**: 301–318.

Hohenberg H., Mannweiler K. and Müller M. (1994) High-pressure freezing of cell suspensions in cellulose capillary tubes. *J. Micros.* **175**: 34–43.

Hollenberg, M. J. and Erickson, A.M. (1973) The scanning electron microscope: potential usefulness to biologists. *J. Histochem. Cytochem.* **21**: 109–130.

Holt, S.C. and Beveridge, T.J. (1982) Electron microscopy: its development and application to microbiology. *Can. J. Microbiol.* **28**: 1–53.

Holzenburg, A. (1988) Preparation of two-dimensional arrays of soluble proteins as demonstrated for bacterial D-ribulose-1,5-bisphosphate carboxylase/oxygenase. In: *Methods in Microbiology*, Vol. 20 (ed. F. Mayer). Academic Press, London, pp. 341–356.

Holzenburg, A. (1995) Electron microscopical analysis of ion channels. In: *Ion Channels – A Practical Approach* (ed. R.H. Ashley). IRL Press, Oxford, pp. 269–290.

Holzenburg, A. (1997) Membrane proteins solved by electron microscopy and electron diffraction. In: *Electron Crystallography*, NATO ASI Series E Vol. 347 (eds. D.L. Dorset, S. Hovmöller and X. Zou) Kluwer, Dordrecht, pp. 323–342.

Holzenburg, A. and Mayer, F. (1989) D-Ribulose-1,5-bisphosphate carboxylase/oxygenase: function-dependent structural changes. *Elec. Micros. Rev.* **2**: 139–169.

Holzenburg, A., Mayer, F., Harauz, G., van Heel, M., Tokuoka, R., Ishida, T., Harata, K., Pal, G.P. and Saenger, W. (1987) Structure of D-ribulose-1,5-bisphosphate carboxylase/oxygenase from *Alcaligenes eutrophus* H16. *Nature* **325**: 730–732.

Holzenburg, A., Wilson, F.H., Finbow, M.E. and Ford, R.C. (1992) Structural investigations of membrane proteins: The versatility of electron microscopy. *Biochem. Soc. Trans.* **20**: 591–597.

Hoppert, M. and Mayer, F. (1995) Electron microscopy technique for immunocytochemical localization of enzymes in methanogenic archaea. In: *Archaea — A Laboratory Manual* (eds F.T. Robb, K.R. Sowers, S. DasSharma, A.R. Place, H.J. Schreier and E.M. Fleischmann). Cold Spring Harbor Press, Cold Spring Harbor, NY, pp. 269–278.

Hoppert, M., Mahoney, T.J., Mayer, F. and Miller, D.J. (1995) Quaternary structure of the hydroxylamine oxidoreductase from *Nitrosomonas europaea. Arch. Microbiol.* **163**: 300–306.

Horisberger, M. (1985) The gold method as applied to lectin cytochemistry in transmission and scanning electron microscopy. In: *Techniques in Immunocytochemistry*, Vol. 3 (eds G.R. Bullock and P. Petrusz). Academic Press, London, pp. 155–178.

Hotta, Y., Kato, H. and Watari, N. (1990) A simple and rapid maceration method for scanning electron microscopy using microwave. *J. Elec. Micros.* **39**: 63–66.

Houwink, A.L. (1953) A macromolecular mono-layer in the cell wall of *Spirillum spec. Biochim. Biophys. Acta* **10**: 360–366.

Hovmöller, S. (1992) Crisp: crystallographic image processing on a personal computer. *Ultramicroscopy* **41**: 121–135.

Huber, R., Wilharm, T., Huber, D., Trincone, A., Burggraf, S., König, H., Rachel, R., Rockinger, I., Fricke, H. and Stetter, K.O. (1992) *Aquifex pyrophilus* gen. nov. sp. nov., represents a novel group of marine hyperthermophilic hydrogen-oxidizing bacteria. *System. Appl. Microbiol.* **15**: 340–351.

Huber, R., Stöhr, J., Hohenhaus, S., Rachel, R., Burggraf, S., Jannasch, H.W. and Stetter, K.O. (1995) *Thermococcus chitonophagus*, sp. nov., a novel, chitin-degrading hyperthermophilic archaeum from a deep-sea hydrothermal vent environment. *Arch. Microbiol.* **164**: 255–264.

Kessels, M.M., Qualmann, B., Klobasa, F. and Sierralta, W.D. (1996) Immunocytochemistry by electron spectroscopic imaging using a homogeneously boronated peptide. *Cell Tissue Res.* **284**: 239–245.

Kistler, J., Aebi, U. and Kellenberger, E. (1977) Freeze-drying and shadowing a two-dimensional periodic specimen. *J. Ultrastruct. Res.* **59**: 76–86.

Knoll, G., Braun, C. and Plattner, H. (1991) Quenched flow analysis of exocytosis in *Paramecium* cells: time course, changes in membrane structure, and calcium requirements revealed after rapid mixing and rapid freezing of intact cells. *J. Cell Biol.* **113**: 1295–1304.

Kölbel, H.K. (1976) Kohleträgerfilme für die hochauflösende Elektronenmikroskopie. Verbesserung von Eigenschaften und Herstellungstechnik. *Mikroskopie* **32**: 1–16.

Körtje, K.H., Paulus, U., Ibsch, M. and Rahmann, H. (1996) Imaging of thick sections of nervous tissue with energy-filtering transmission electron microscopy. *J. Micros.* **183**: 89–101.

Kushida, H. (1967) A new embedding method employing DER 736 and Epon 812. J. *Elec. Micros.* **16**: 278–283.

Labhart, P. and Koller, T. (1981) Electron microscopic specimen preparation of rat liver chromatin by a modified Miller spreading technique. *Eur. J. Cell Biol.* **24**: 309–316.

Lang, D. and Mitani, M. (1970) Simplified quantitative electron microscopy of biopolymers. *Biopolymers* **9**: 373–379.

Lanzavecchia, S., Bellon, P.L. and Scatturin, V. (1993) SPARK, a kernel of software programs for spacial reconstruction in electron microscopy. *J. Micros.* **171**: 255–266

Lebermann, R. (1965) Use of uranyl formate as negative stain. *J. Mol. Biol.* **13**: 606.

Leduc, E.H. and Bernhard, W. (1967) Recent modifications of glycol methacrylate embedding procedure. *J. Ultrastruct. Res.* **19**: 196–199F.

Lennard, P. (1990) Image analysis for all. *Nature* **347**: 103–104.

Loud, A.V. and Anversa, P. (1984) Biology of disease — morphometric analysis of biologic processes. *Lab. Invest.* **50**: 250–261.

Luft, J.H. (1956) Permanganate — a new fixative for electron microscopy. *J. Biophys. Biochem. Cytol.* **2**: 799–801.

Luft, J.H. (1959) The use of acrolein as a fixative for light and electron microscopy. *Anat. Rec.* **133**: 305.

Lünsdorf, H. and Spiess, E. (1987) A rapid method of preparing perforated supporting foils for the thin carbon films used in high resolution transmission electron microscopy. *J. Micros.* **144**: 211–213.

Markham, R., Frey, S. and Hills, G.J. (1963) Methods for the enhancement of image detail and accentuation of structure in electron microscopy. *Virology* **20**: 88–102.

Martin, R. (1996) The structure of the neurofilament cytoskeleton in the squid giant axon and synapse. *J. Neurocytol.* **25**: 547–554.

Massover, W.H. and Marsh, P. (1997) Unconventional negative stains: heavy metals are not required for negative staining. *Ultramicroscopy* **69**: 139–150.

Mayer, F. (1986) Cytology and morphogenesis of bacteria. In: *Encyclopedia of Plant Anatomy*, Vol. 6, Part 2 (eds H.J. Braun, S. Carlquist, P. Ozenda and I. Roth). Borntraeger, Berlin.

Mayer, F. and Friedrich, C.G. (1986) Higher order structural organization of the nucleoid in the thiobacterium *Thiosphaera pantotropha*. *FEMS Microbiol. Lett.* **73**: 109–112.

Mayer, F. and Hoppert, M. (1997) Determination of the thickness of the boundary layer surrounding bacterial PHA inclusion bodies, and implications for models describing the molecular architecture of this layer. *J. Basic Microbiol.* **37**: 45–52.

Mayer, F. and Rohde, M. (1988) Analysis of dimensions and structural organization of proteoliposomes. In: *Methods in Microbiology*, Vol. 20 (ed. F. Mayer). Academic Press, London, pp. 283–292.

Meek, G.A. (1976) *Practical Electron Microscopy for Biologists* (2nd Edn). John Wiley & Sons, London.

Mellema, J.E., Van Bruggen, E.F.J. and Gruber, M. (1967) Uranyl oxalate as negative stain for electron microscopy of proteins. *Biochim. Biophys. Acta* **140**: 180–182.

Miller, S.E. and Howell, D.N. (1997) Concerted use of immunologic and ultrastructural analyses in diagnostic medicine: immunoelectron microscopy and correlative microscopy. *Immunol. Invest.* **26**: 29–38.

Millonig, G. (1961) Advantages of a phosphate buffer for OsO_4 solutions in fixation. *J. Appl. Phys.* **32**: 1637.

Misell, D.L. (1978) Image analysis, enhancement and interpretation. In: *Practical Methods in Electron Microscopy*, Vol. 7 (ed. A. M. Glauert). Elsevier, Amsterdam.

Misell, D.L. and Brown, E.B. (1987) Electron diffraction: an introduction for biologists. In: *Practical Methods in Electron Microscopy*, Vol. 12 (ed. A. M. Glauert). Elsevier, Amsterdam.

Monosov, E.Z., Wenzel, T.J., Luers, G.H., Heyman, J.A. and Subramani, S. (1996) Labeling of peroxisomes with green fluorescent protein in living *P. pastoris* cells. *J. Histochem. Cytochem.* **44**: 581–589.

Moor, H. (1973) Evaporation and electron guns. In: *Freeze-etching, Techniques and Applications* (eds E.L. Benedetti and P. Favard). Soc. Française de Microscopie Électronique, Paris, p. 27.

Müller, M., Koller, T. and Moor, H. (1979) Preparation and use of aluminium films for high resolution electron microscopy of macromolecules. *Proc. 7th Int. Congr. Elec. Micros. Grenoble* **1**: 633–634.

Muscatello, U. and Horne, R.W. (1968) Effect of the tonicity of some negative-staining solutions on elementary structure of membrane-bounded systems. *J. Ultrastruct. Res.* 25: 73–83.

Newman, G.R. and Hobot, J.A. (1987) Modern acrylics for post-embedding immunostaining techniques. *J. Histochem. Cytochem.* **35**: 971–981.

Olins, A.L., Olins, D.E., Levy, H.A., Shah, M.B. and Bazett-Jones, D.P. (1993) Electron microscope tomography of Balbiani ring hnRNP substructure. *Chromosoma* **102**: 137–144.

Olins, A.L., Olins, D.E., Olman, V., Levy, H.A. and Bazett-Jones, D.P. (1994) Modeling the 3-D RNA distribution in the Balbiani ring granule. *Chromosoma* **103**: 302–310.

Osmani, S.A., Mayer, F., Marston, F.A.O., Selmes, I.P. and Scrutton, M.C. (1984) Pyruvate carboxylase from *Aspergillus nidulans*: effects of regulatory modifiers on the structure of the enzyme. *Eur. J. Biochem.* **139**: 509–518.

Packter, N.M. and Olukoshi, E.R. (1995) Ultrastructural studies of neutral lipid localisation in *Streptomyces*. *Arch. Microbiol.* **164**: 420–427.

Paul, T.R. and Beveridge, T.J. (1993) Ultrastructure of mycobacterial surfaces by freeze-substitution. *Int. J. Med. Microbiol. Virol. Parasitol. Infect. Dis.* **279**: 450–457.

Plattner, H. and Bachmann, L. (1982) Cryofixation: a tool in biological ultrastructural research. *Int. Rev. Cytol.* **79**: 237–304.

Qualmann, B., Kessels, M.M., Klobasa, F., Jungblut, P.W. and Sierralta, W.D. (1996) Electron spectroscopic imaging of antigens by reaction with boronated antibodies. *J. Micros.* **183**: 69–77.

Read, N.D. and Jeffree, C.E. (1991) Low temperature scanning electron microscopy in biology. *J. Micros.* **161**: 59–72.

Reimer, L. (1978) Scanning electron microscopy — present state and trends. *Scanning* **1**: 3–16.

Reynolds, E.S. (1963) The use of lead citrate at high pH as an electron-opaque stain in electron microscopy. *J. Cell Biol.* **17**: 208–213.

Rieger, G., Rachel, R., Hermann, R. and Stetter, K.O. (1995) Ultrastructure of the hyperthermophilic archaeon *Pyrodictium abyssi*. *J. Struct. Biol.* **115**: 78–87.

Rieger, G., Müller, K., Hermann, R., Stetter, K.-O. and Rachel, R. (1997) Cultivation of hyperthermophilic archaea in capillary tubes resulting in improved preservation of fine structures. *Arch. Microbiol.* **168**: 373–379.

Robards, A.W. and Sleytr, U.B. (1985) in *Practical Methods in Electron Microscopy*, Vol. 10: *Low Temperature Methods in Biological Electron Microscopy* (ed. A.M. Glauert). Elsevier, Amsterdam.

Roberts, I.M. (1975) Tungsten coating — a method of improving glass microtome knives for cutting ultrathin sections. *J. Micros.* **103**: 113–119.

Robinson, D.G., Ehlers, U., Herken, R., Herrmann, B., Mayer, F. and Schürmann, F.W. (1987) *Methods of Preparation for Electron Microscopy*. Springer-Verlag, Berlin.

Rochaix, J.D. and Malnoe, P. (1978) Anatomy of the chloroplast ribosomal DNA of *Chlamydomonas reinhardii*. *Cell* **15**: 661–670.

Rohde, M., Gerberding, H., Mund, T. and Kohring, G.W. (1988) Immunoelectron microscopic localization of bacterial enzymes: pre- and postembedding labeling techniques on resin-embedded samples. In: *Methods in Microbiology*, Vol. 20 (ed. F. Mayer). Academic Press, London, pp. 175–210.

Roos, N. and Morgan, A.J. (1990) Cryopreparation of thin biological specimens for electron microscopy: Methods and applications. *Microscopy Handbooks*, Vol. 21. Royal Microscopic Society and Oxford University Press, Oxford.

Roth, J., Bendayan, M. and Orci, L. (1978) Ultrastructural localization of intracellular antigens by the use of Protein A–gold complex. *J. Histochem. Cytochem.* **26**: 1074–1081.

Roth, J., Bendayan, M. and Orci, L. (1980) FITC–Protein A–Gold complex for light and electron microscopic immunocytochemistry. *J. Histochem. Cytochem.* **28**: 55–57.

Sabatini, D.D., Bensch, K. and Barrnett, R.J. (1963) Cytochemistry and electron microscopy: the preservation of cellular structure and enzymatic activity by aldehyde fixation. *J. Cell Biol.* **17**: 19–58.

Sakata, S., Hotsumi, S. and Watanabe, H. (1991) One nanometer thickness specimen supporting film. *J. Elec. Micros.* **40**: 67–69.

Schmid, M.F., Dargahi, R. and Tam, M.W. (1993) Spectra: a system for processing electron images of crystals. *Ultramicroscopy* **48**: 251–264.

Scott, J.E. (1972) The histochemistry of alcian blue. *Histochemie* **29**: 129–133.

Shires, M., Goode, N.P., Crellin, D.M. and Davidson, A.M. (1990) Immunogold–silver staining of mesangial antigen in Lowicryl K4M- and LR gold-embedded renal tissue using epipolarization microscopy. *J. Histochem. Cytochem.* **38**: 287–289.

Shotton, D.M. and Severs, N.J. (1995) An introduction to freeze fracture and deep etching. In: *Rapid Freezing, Freeze, Fracture and Deep Etching* (eds N.J. Severs and D.M. Shotton). Wiley–Liss, New York, pp. 1–30.

Sitte, H. (1984) Equipment for cryofixation, cryoultramicrotomy and cryosubstitution in bio-medical TEM routines. *Zeiss Information (Mag. Elec. Micros.)* **3**: 25–31.

Sjöstrand, F.S. (1990) Common sense in electron microscopy. *J. Struct. Biol.* **103**: 135–139.

Skaer, H. le B. (1982) Chemical cryoprotection for structural studies. *J. Micros.* **125**: 137–147.

Sleytr, U.B. and Messner, P. (1983) Crystalline surface layers on bacteria. *Ann. Rev. Microbiol.* **37**: 311–339.

Sleytr, U.B. and Robards A.W. (1977) Plastic deformation during freeze-cleaving: a review. *J. Micros.* **110**: 1–25.

Sleytr, U.B. and Robards, A.W. (1981) Understanding the artefact problem in freeze-fracture replication: a review. *J. Micros.* **101**: 187–195.

Spiess, E. and Lurz, R. (1988) Electron microscopic analysis of nucleic acids and nucleic acid–protein complexes. In: *Methods in Microbiology*, Vol. 20 (ed. F. Mayer). Academic Press, London, pp. 293–323.

Springer, E.L. and Roth, I.L. (1972) Scanning electron microscopy of bacterial colonies. I. *Diplococcus pneumoniae* and *Streptococcus pyogenes*. *Can. J. Microbiol.* **18**: 219–223.

Spurr, A.R. (1969) A low-viscosity epoxy resin embedding medium for electron microscopy. *J. Ultrastruct. Res.* **26**: 31–43.

Stuart, M. (1991) in *Electron Microscopy in Biology — A Practical Approach* (ed. R. Harris). IRL Press, Oxford, pp. 229–242.

Stupperich, E., Juza, A., Hoppert, M. and Mayer, F. (1993) Cloning, sequencing and immunological characterization of the corrinoid-containing subunit of the N^5-methyltetrahydromethanopterin:coenzyme-M methyltransferase from *Methanobacterium thermoautotrophicum*. *Eur. J. Biochem.* **217**: 115–121.

Tauschel, H.-D. (1988) Localization of bacterial enzymes by electron microscopic cytochemistry as demonstrated for the polar organelle. In: *Methods in Microbiology*, Vol. 20 (ed. F. Mayer). Academic Press, London, pp. 237–259.

Tokuyasu, K.T. (1986) A technique for ultramicrotomy of cell suspensions and tissues. *J. Cell Biol.* **57**: 551–565.

Tvedt, K.E., Kopstadt, G. and Haugen, O.A. (1984) A section press and low elemental support for enhanced preparation of freeze-dried cryosections. *J. Micros.* **133**: 285–290.

Tyler, J.M. and Branton, D. (1980) Rotary shadowing of extended molecules dried from glycerol. *J. Ultrastruct. Res.* **71**: 95–102.

Umrath, W. (1974) Cooling bath for rapid freezing in electron microscopy. *J. Micros.* **101**: 103–105.

Umrath, W. (1983) Calculation of the freeze-drying time for electron microscopical preparations. *Mikroskopie* **40**: 9–34.

Valentine, R.C., Shapiro, B.M. and Stadtman, E.R. (1968) Regulation of glutamine synthetase. XII. Electron microscopy of the enzyme from *Escherichia coli*. *Biochemistry* **7**: 2143–2152.

Van Bruggen, E.F.J., Wiebinga, E.H. and Gruber, M. (1960) Negative-staining electron microscopy of proteins at pH values below their isoelectric points. Its application to hemocyanin. *Biochim. Biophys. Acta* **42**: 171–172.

Van Bruggen, E.F.J., Wiebinga, E.H. and Gruber, M. (1962) Structure and properties of hemocyanins. I. Electron micrographs of hemocyanin and apohemocyanin from *Helix pomatia* at different pH-values. *J. Mol. Biol.* **4**: 1–7.

Van Harrefeld, A. and Crowell, J. (1964) Electron microscopy after rapid freezing on a metal surface and substitution fixation. *Anat. Rec.* **149**: 381–385.

Van Harrefeld, A., Trubatch, J. and Steiner, J. (1974) Rapid freezing and electron microscopy for the arrest of physiological processes. *J. Micros.* **100**: 189–198.

Varga, A.R. and Staehelin, L.A. (1983) Spatial differentiation in photosynthetic and non-photosynthetic membranes of *Rhodopseudomonas palustris*. *J. Bacteriol.* **154**: 1414–1430.

Venable, J.H. and Coggeshall, R. (1965) A simplified lead citrate stain for use in electron microscopy. *J. Cell. Biol.* **25**: 407–408.

Vogt, B., Berker, R. and Mayer, F. (1995) Improved contrast by a simplified post-staining procedure for ultrathin sections of resin-embedded bacterial cells: Application of ruthenium-red. *J. Basic Microbiol.* **35**: 349–355.

Vollenweider, H.J., Sogo, J.M. and Koller, T. (1975) A routine method for protein-free spreading of double- and single-stranded nucleic acid protein molecules. *Proc. Natl Acad. Sci. USA* **72**: 83–87.

Walter-Mauruschat, A., Aragno, M., Mayer, F. and Schlegel, H.G. (1977) Micromorphology of Gram-negative hydrogen bacteria. II. Cell envelope, membranes, and cytoplasmic inclusions. *Arch. Microbiol.* **114**: 101–110.

Walther P., Wehrli E., Hermann R. and Müller, M. (1995) A coating and imaging technique to reduce the effects of beam damage for high-resolution low-temperature SEM. *J. Micros.* **179**: 229–237.

Watson, L.P., McKee, A.E. and Merrell, B.R. (1980) Preparation of microbial specimens for electron microscopy. In: *Scanning Electron Microscopy*, Vol. 1980/II (ed. E.D. Johari). Scanning Electron Microscopy Inc., Chicago, pp. 45–56.

Westmoreland, B., Szybalski, W. and Ris, H. (1969) Mapping of deletions and substitutions in heteroduplex DNA molecules of bacteriophage Lambda by electron microscopy. *Science* **163**: 1343–1348.

Wohlrab, F. and Gossrau, R. (1992) *Katalytische Enzymhistochemie*. G. Fischer Verlag, Jena.

Wurtz, M. (1992) Bacteriophage structure. *Elec. Micros. Rev.* **5**: 283–309.

Zingsheim, H.P., Abermann, R. and Bachmann, L. (1970) Apparatus for ultrashadowing of freeze-etched electron microscopic specimens. *J. Phys. E: J. Sci. Instrum.* **3**: 39–42.

General textbooks on methods for electron microscopy

Aldrich, H.C. and Todd, W.J. (1986) *Ultrastructure Techniques for Microorganisms.* Plenum Press, New York.

Glauert, A.M. (1973–1987) *Practical Methods in Electron Microscopy*, Vol. 1–12. Elsevier, Amsterdam.

Hayat, M.A. (1975) *Positive Staining for Electron Microscopy.* Van Nostrand Reinhold, New York.

Hayat, M.A. (1978) *Introduction to Biological Scanning Electron Microscopy.* University Park Press, Baltimore.

Kay, D. (1965) *Techniques for Electron Microscopy.* Blackwell Scientific Publications, Oxford.

Mayer, F. (1988) *Methods in Microbiology*, Vol 20: *Electron Microscopy in Microbiology.* Academic Press, London.

Robinson, D.G., Ehlers, U., Herken, R., Herrmann, B., Mayer, F. and Schürmann, F.W. (1987) *Methods of Preparation for Electron Microscopy.* Springer-Verlag, Berlin.

Appendix B
Suppliers

Accessories and consumables in electron microscopy

If not indicated otherwise, the suppliers included in this list provide most, if not all accessories and consumables for EM preparations.

Agar Scientific Ltd, 66a Cambridge Road, Stansted, Essex CM24 8DA, UK.

Aurion, 6702 AA Wageningen, The Netherlands (immunogold reagents).

British BioCell, Golden Gate, Ty Glas Avenue, Cardiff CF4 5DX, UK (immunogold reagents).

Delaware Diamond Knives Inc., 3825 Lancaster Pike, Wilmington, DE 19805, USA (diamond knives).

Drukker International BV, Beversestraat 20, 5431 SH Cuijk, The Netherlands (diamond knives).

Edmund Scientific Co., 101 East Glouchester Pike, Barrington, NJ 08007-1380, USA (scale loupes).

Electron Microscopy Sciences, P.O. Box 251, Fort Washington, PA 19034, USA.

Ernest F. Fullam Inc., 900 Albany Shaker Road, Latham, NY 12110, USA.

Goldmark Biologicals, 437 Lock Street, Phillipsburg, NJ08865, USA (immunogold reagents).

Goodfellow Cambridge Ltd, Cambridge Science Park, Cambridge CB4 4DJ, UK.

and

Goodfellow Corporation, 800 Lancaster Avenue, Berwyn, PA 19312-1780, USA (specialist supplier of small quantities of materials for R&D, prototype work and design; for SEM work, the self-supporting carbon sheets are of particular interest).

J.B. EM Services Inc., P.O. Box 693, Pointe-Claire, Dorval, Quebec H9R 4S8, Canada.

London Resin Co. Ltd, P.O. Box 34, Basingstroke, Hants RG25 2EX, UK (LR White, LR Gold embedding resins).

Mallinckrodt Laboratory Chemicals, 222 Red School Lane, Phillipsburg, NJ 08865, USA (tannic acid for preparation of colloidal gold).

Media Cybernetics, L.P., 8484 Georgia Avenue, Silver Spring, MD 20910, USA (image analysis software).

Nanoprobes, 25 East Loop Road, Suite 113, Stony Brook, NY 11790-3350, USA (1 nm gold colloids and other immunogold reagents).

Pelco International, P.O. Box 492477, Redding, CA 96049-2477, USA.

Polysciences Inc., Paul Valley Industrial Park, Warrington, PA 18976, USA (chemicals, resin embedding kits).

Science Services Ltd, Greenwood House, 4/7 Salisbury Court, London EC4 8BT, UK.

Scott Scientific Inc., P.O. Box 66552, Station Cavendish, Montreal, Quebec, H4W 3J6, Canada.

TAAB Ltd, 3 Minerva Park, Aldermaston, Berks RG7 4QW, UK.

Technical equipment

BAL-TEC Inc., Fl-9496 Balzers, Principality of Liechtenstein (critical-point dryers, sputter coaters, vacuum evaporators, vacuum pumps, equipment for deep temperature embedding, freeze-drying and freeze-etching).

Cressington Scientific Instruments Ltd, 34 Chalk Hill, Watford WD1 4BX, UK (benchtop coating and sputtering units, high-resolution sputtering and coating, freeze-fracture/etching units).

Denton Vacuum Inc., Cherry Hill Industrial Center, Cherry Hill, NJ 08003, USA (vacuum pumps, coating units).

Edwards High Vacuum International, Manor Royal, Crawley, W. Sussex RH10 2LW, UK (vacuum pumps, coating units).

GATAN Inc., 780 Commonwealth Drive, Warrendale, PA 15086, USA (cameras, energy-filtered imaging, specimen preparation, SEM digital imaging, specimen holders).

Oxford Instruments, Old Station Way, Eynsham, Witney, Oxon OX8 1TL, UK (cryo-specimen holders, cryo-transfer workstations, anticontaminators, SEM cryopreparation chambers, temperature control units).

Reichert-Cambridge Instruments, Hernalser Hauptstr. 219, A-1171 Wien, Austria (ultramicrotomes, instrumentation for cryopreparations).

Scott Scientific Inc., PO Box 66552, Station Cavendish, Montreal, Quebec, H4W 3J6, Canada (imaging systems, photographic equipment, tissue sectioning equipment, critical-point dryers, sputter coaters, vacuum evaporators).

Appendix C

Useful Internet resources

It is impossible to list all of the most interesting and rapid changing "electron microscopy" Internet pages. Numerous institutions offer insight in their image archives, preparation techniques (a comprehensive manual is provided by the Center for Cell Imaging at the Yale University School of Medicine: http://info.med.yale.edu/cellimg), equipment and literature. Three sources indexing these pages are given:

http://www.mwrn.com/guide.htm
http://www.ou.edu/research/electron/www-vl/
http://cimewww.epfl.ch/emyp/

Appendix D

Hazardous chemicals in electron microscopic preparation techniques

As a general rule, the experimentor should know, before starting an experiment, all data concerning potential risks of inflammability, health damage and damage to the environment. The short advice given here refers to chemicals commonly used in numerous preparation techniques and is restricted to some of the most important. Extended information, if not found in chemical literature, *must* be provided by the respective supplier.

Aldehydes: These are toxic after contact to skin, eyes and mucosa. Avoid exposure to vapours and paraformaldehyde powder. Aldehydes may cause irreversible health damage.

Resins: Nanoplast releases formaldehyde during condensation. The component *p*-toluenesulphonic acid irritates skin, eyes and respiratory tract by direct contact. Methacrylates, similar to other embedding media, causes eczema and allergic reactions. Even short exposure of skin to Lowicryl resin causes eczema at the contact site. Avoid contact with skin and eyes, avoid inhaling the vapours from the resins. Use a well-ventilated fume hood or a place with efficient ventilation for all procedures with liquid resin; wear vinyl gloves (vapours will pass through latex gloves).

Sigmacote: Causes burns, and harmful if swallowed, inhaled or absorbed through skin. Do not get in eyes, on skin or clothing, and avoid breathing vapour. Wear gloves and work in a fume hood.

Tetrachloroauric acid: Irritating to eyes, respiratory system and skin. Wear gloves and work in a fume hood.

Heavy metal salts and their solutions (uranyl acetate, lead citrate, osmium tetroxide, cacodylate buffer): Very toxic by inhalation or if swallowed; uranyl acetate is radioactive; osmium tetroxide is a strong oxidizing agent and highly reactive. Danger of cumulative effects. Avoid breathing the powder; handle the powder with extreme care in a fume hood, and wear gloves.

Consult the local safety office for further guidance in the appropriate use of radioactive materials.

Cryogens: Liquid nitrogen must be stored in appropriate Dewar containers in well-ventilated places to avoid enrichment of nitrogen in the gaseous atmosphere. Liquid nitrogen causes severe frostbite when in direct contact with skin for longer than a few seconds. Therefore, it is very dangerous to handle nitrogen in clothing, such as rubber boots, where nitrogen can accumulate. Protective eyeglasses should allow any spilled liquid to flow through them instead of accumulating behind the eyeglasses.

Boiling of mixtures of liquid and solid nitrogen may lead to sudden bursts of nitrogen.

Liquefaction of the explosive gases propane and ethane must be performed in well-ventilated places. Specialized burners are available from suppliers for cryo-devices to burn the cryogen after use.

Index